Protected Areas, National Parks and Sustainable Future

Edited by Ahmad Naqiyuddin Bakar and Mohd Nazip Suratman

Published in London, United Kingdom

IntechOpen

Supporting open minds since 2005

Protected Areas, National Parks and Sustainable Future
http://dx.doi.org/10.5772/intechopen.77900
Edited by Ahmad Naqiyuddin Bakar and Mohd Nazip Suratman

Contributors
Joel Heinen, Nabin Baral, Prakash Paudel, Jay Sah, Adil - Siswanto, Djumilah Hadiwidjojo, Christian Hébert, Eric Domaine, Louis Bélanger, Nguyen Kim Dung, Patrick O. Waeber, H. Andry Rasamuel, Serge C. Rafanoharana, Lucienne Wilmé, Lais Dib, João Pedro Palmer, Camila Lima, Otilio Bastos, Claudia Uchôa, Maria Regina Amendoeira, Augusto Bastos, Alynne Barbosa, Ahmad Bakar

First published in London, United Kingdom, 2020 by IntechOpen
IntechOpen is the global imprint of INTECHOPEN LIMITED, registered in England and Wales, registration number: 11086078, 7th floor, 10 Lower Thames Street, London, EC3R 6AF, United Kingdom
Printed in Croatia

British Library Cataloguing-in-Publication Data
A catalogue record for this book is available from the British Library

Additional hard and PDF copies can be obtained from orders@intechopen.com

Protected Areas, National Parks and Sustainable Future
Edited by Ahmad Naqiyuddin Bakar and Mohd Nazip Suratman
p. cm.
Print ISBN 978-1-78984-229-6
Online ISBN 978-1-78984-230-2
eBook (PDF) ISBN 978-1-83968-372-5

For EU product safety concerns:
IN TECH d.o.o., Prolaz Marije Krucifikse Kozulić 3, 51000 Rijeka, Croatia, info@intechopen.com or visit our website at intechopen.com.

Meet the editors

Ahmad Naqiyuddin Bakar, CMILT (UiTM, IIUM, Exeter, Hull), is an associate professor at the Faculty of Administrative Science and Policy Studies, Universiti Teknologi MARA (UiTM), Malaysia. He is currently rector of the UiTM Johor branch. Prior to this, he was Head of the Research and Publication Unit at the Malaysia Institute of Transport after having previously served as Head of Institutional Research under the Strategic Planning and Information (CSPI) Centre at the Office of the Deputy Vice-Chancellor (Research and Innovation) of UiTM. He is also a senior research fellow at the Centre for Biodiversity and Sustainable Development, UiTM. He holds a Professional Certificate in Science, Technology and Innovation (STI) in Policy Management. Much of his research work is focused on public policy, sustainable community development and biodiversity, and human capital management.

Mohd Nazip Suratman is a professor of forestry in the Faculty of Applied Sciences, Universiti Teknologi MARA (UiTM), Malaysia. He earned a BSc degree in Forestry from the Universiti Putra Malaysia, an MSc from the University of Nebraska-Lincoln, USA, and a PhD from the University of British Columbia, Canada, specializing in remote sensing for forest resources management. His current research interests include biodiversity conservation and combining remote sensing and forest inventory for biomass modeling. Throughout his academic career, he has acquired many national and international research grants and carried out a number of research projects that have led to publications in leading journals, books, and chapters in books. He was the recipient of the best researcher award of UiTM for 2015 in recognition of his outstanding research work in science, technology, and medicine cluster. He has been the Head of the Centre for Biodiversity and Sustainable Development and the Deputy Vice-Chancellor for Research and Innovation at UiTM, Malaysia, since 2018.

Contents

Preface

National parks and protected areas are earth's most diversified ecosystems and form the basis of life on earth because all living organisms are supported by the interactions among them. This book explores the context and practice of national parks with countries' obligations to safeguard biodiversity through the protection and management of forest-protected areas. The book examines the wider impacts of national parks within the scope of an integrated environmental hub at the global and regional level and eventually delves into the country case.

The book covers three areas: theoretical underpinnings and concepts relating to national parks, exploring their various modalities and integrated concerns for the environment; an empirical review in lieu of effective management of protected areas as defined by the World Conservation Union IUCN, addressing the efficient use of human and material resources, including national/agency-protected area regulations and legislation, policies, international conventions and designations, management plans, and/or agreements associated with specific identified areas; and evaluation of challenges underlying each country's intention to gauge the potentials of a national park and pinpoint adequate attention on exploiting new strategies for national parks and protected area management.

This book is a compilation of research papers illustrating the immense diversity and value of national parks and protected areas through a case reference of selected countries in Asia, Africa, and North America.

It is hoped that this book will contribute to harnessing any synergies and complementarities by learning from the experience of both developed and developing countries. Evidence shows that through properly designed and implemented policies and instruments, it can significantly increase the impact on initiatives that conserve national parks and protected areas, thereby discouraging behaviors that are detrimental to species and ecosystems.

Writing a book is harder than we thought but more rewarding than we could have ever imagined. None of this would have been possible without the understanding and the true friendship that we both share. We stood aside during every struggle and faced the challenges together. We are eternally grateful to the great team at IntechOpen. Mr. Gordan Tot and the team have taught us discipline, empathy, manners, respect, and so much more, which has motivated us to complete the project successfully. We truly appreciate the fact that we would not have been able to complete the project if IntechOpen hadn't given us generous time, which we desperately needed. Although shouldering great responsibilities and commitments with our university's administrative duties, our time for research and writing was worth it. To everyone at the UiTM who enabled us to be the top executive management that we are honored to be a part of, thank you for letting us serve, for being a part of our amazing academia, and for the trust given to us to turn their continue our core duties as researchers.

Thanks go our families, particularly our wives Rafidah and Zara, for always being the persons we could turn to during tough times. They obediently sustained us in ways that we never knew that we needed. Sincere appreciation also goes to our children. So thankful to have all of you back in our lives.

Finally, to all those colleagues who have been providing us with assistance directly and indirectly: UiTM's Faculty of Administrative Science and Policy Studies, Faculty of Applied Science, Malaysia Institute of Transport, the Centre for Biodiversity and Sustainable Development, UiTM, the Chancellery, UiTM Johor Branch, and last but not least, the Research Management Institute, UiTM.

Dr. Ahmad Naqiyuddin Bakar
Faculty of Administrative Science and Policy Studies,
UiTM,
Shah Alam, Selangor, Malaysia

Malaysia Institute of Transport,
UiTM,
Shah Alam, Selangor, Malaysia

Centre for Biodiversity and Sustainable Development,
UiTM,
Puncak Alam, Selangor, Malaysia

UiTM Johor Branch,
Segamat Campus, Segamat, Johor, Malaysia

Dr. Mohd Nazip Suratman
Centre for Biodiversity and Sustainable Development,
UiTM,
Puncak Alam, Selangor, Malaysia

Faculty of Applied Science,
UiTM,
Shah Alam, Selangor, Malaysia

Chapter 1

Introductory Chapter: Today's National Parks (NPs) and Protected Areas (PAs) for a Sustainable Future

Ahmad Naqiyuddin Bakar

1. Introduction

The decline in the world's biodiversity is recognised as a key facet of the environmental crisis in which the twentieth century has seen the most far-reaching ecological change that provides the greatest threat to the planet and to human survival. There is continuing understanding, starting from the famous Earth Summit in 1992 through to the 2012 Rio Conference, that urgent actions are needed to secure core natural assets such as woodlands and biodiversity matters. At the turn of the century, world leaders adopted the United Nations Millennium Declaration and outlined eight chapters and key objectives or targets and 2015 as the deadline [1]. These targets are globally known as millennium development goals—ranging from peace; development; environment; human rights; and the vulnerable, hungry, and poor. In 2015, countries adopted the 2030 Agenda for Sustainable Development and its 17 Sustainable Development Goals. Implementation and success will rely on countries' own sustainable development policies, plans, and programmes and will be led by countries. The Sustainable Development Goals (SDGs) will be a compass for aligning countries' plans with their global commitments [2].

"As we hurdle towards 2030, Sustainable Development Goals 14 and 15 cannot be overlooked, because that's where biodiversity and nature reside—life below water, 14, and life on land, 15. And to get to 2030 we need to get biodiversity right," said IUCN Director General Inger Andersen [3]. The Protected Planet Report 2018 reviews the achievement of Aichi Biodiversity Target 11, which stresses for the effective and equitable management of 17% of terrestrial and 10% of coastal and marine areas by 2020. The report confirmed that the world is well in line to meet the coverage aspect of Target 11 and highlighted the need to meet related aspects by 2020 [4].

This chapter focuses on the management of PAs and NPs—arguably the world's most depleted resources and the initiatives taken by countries when delivering their obligations mainly under the auspices of IUCN both within and beyond national jurisdiction. The chapter falls into five sections. The brief review of the world's biodiversity, the role of IUCN, and the impending threats to world's biodiversity is outlined in Section 1. Next, Section 2 outlines the "IUCN Categories" as well as the growing importance of PAs and NPs. Section 3 presents the major threats to conservation and the implications they brought to PAs and NPs. Section 4 presents a review of some major global initiatives, which exist alongside the statutory designations of the PA system in some countries as case references. All these initiatives have been led by "third sector" organisations, and all attempts have been made to complement the global approach to ecological management with partnership working and public participation. The chapter concludes, in Section 5, that within the present economic and political structures of the PAs and NPs, these new initiatives represent individually imaginative and in aggregate vital adjuncts to areas protected by formal (statutory) designation.

2. Protected areas and national parks: a global commitment

The International Union for Conservation of Nature (IUCN) is an international organisation working in the field of nature conservation and sustainable use of natural resources. It offers public, private, and non-governmental organisations with the "nature-based solutions" to reversing environmental declines and securing a healthy, liveable planet. Since its establishment in 1948, IUCN has evolved into the world's largest and most diverse environmental network. It harnesses the experience, resources, and reach of its offices in more than 160 countries (plus 1300 member organisations) and runs projects all around the world benefited from a network of 13,000+ voluntary scientists and experts spanning the globe [5]. This validates the IUCN as the global authority on the status of the natural world and the measures needed to safeguard it with the world's largest and most diverse environmental network. It continues to champion nature-based solutions as key to the implementation of international agreements such as the Paris Climate Change Agreement and the 2030 Sustainable Development Goals. IUCN experts are organised into six commissions dedicated to species survival, environmental law, protected areas, social and economic policy, ecosystem management, and education and communication [5]. IUCN works across a wide range of themes related to conservation, environmental, and ecological issues, covering business and biodiversity; climate change; ecosystem management; environmental law; forests; gender; global policy; governance and rights; marine and polar; protected areas; science and economics; species; water; and world heritage [6].

2.1 Natural resources at risks

At present, however, biodiversity is facing devastating effects that are directly or indirectly resulting from the rapid growth in human population. According to the Population Division of the Department of Economic and Social Affairs of the United Nations [7], the world's population is predicted to increase to 9.1 billion by 2050. Gross observation shows that most of this growth will occur in developing countries that record human activities and their associated impact to most of the world's biodiversity. It will aggravate to many hazards and risks on the world's ecosystems and species because the majority of human populations derive their livelihoods from the increased susceptibilities of these biological resources. In addition to the increasing overall vulnerabilities of population growth, biodiversity is also threatened by pollution, climate change, habitat loss, and invasive species [8]. The author's purpose is merely to present the current practice of various global initiatives pertaining to protected area and national park in comparative fashion. Since IUCN has been deeply involved with protected areas and national parks from its very beginnings, the roles performed by this key international organisation will also be highlighted. By showing what has been done with such implementation, each initiative illustrates the interaction of international and national agencies as well as NGOs in support of such activities. As we compare and contrast across regions, one should note that countries have their power to ensure that species and ecosystems are maintained as part of the human habitat. This one factor presents unique challenges to sustainability planning implementation, which will be outlined.

Informed estimates are that as many as a million species may be gone forever by the turn of the century [9]. As a matter of fact, of all the species that have ever existed on Earth, 99% have become extinct [10].

Apart from the physical damage plastic causes, it contains harmful chemicals, damaging to both humans and wildlife. Plastic additives such as phthalates and chlorinated, brominated, and fluorinated compounds pose significant risks to human health. In the absence of decisive action, 1.2 million tonnes of additives per year could enter our oceans by 2050, and together with the plastic in soil and freshwater, they will continue to contaminate our food chain and water supplies. The oceans provide protein-rich food for billions of people, so the chemical contamination they introduce into the food chain poses a serious threat to food security and health. Worldwide, marine plastic pollution is costing us between US$ 13 and 40 billion per year as tourism, fisheries, and shipping revenues are lost [3].

2.2 What is a protected area?

The World Conservation Union (IUCN) widely refers protected areas as "... a clearly defined geographical space recognized, dedicated and managed, through legal or other effective means to achieve the long-term conservation of nature with associated ecosystem services and cultural values" [11]. Since Yellowstone Park in USA was first officially protected, in the year 1872, the number and popularity of protected areas are rapidly growing around the world [1].

Protected areas—national parks, wilderness areas, community conserved areas, nature reserves, and so on—are a mainstay of biodiversity conservation, while also contributing to people's livelihoods, particularly at the local level. Protected areas provide us food, clean water supply, medicines, and protection from the impacts of natural disasters and are at the core of efforts involving actors, institutions, strategies, and policies towards conserving nature. It has been estimated that the global network of protected areas stores at least 15% of terrestrial carbon; thereby, their role in helping mitigate and adapt to climate change is also increasingly acknowledged [1].

2.3 IUCN protected area categories system

2.3.1 Protected area categories

Protected areas are classified according to their management objectives set out in the IUCN protected area management categories. The categories set forth a multitude of management strategies aimed at protecting areas, are recognised by international bodies such as the United Nations and by many national governments as the global standard, and as such are increasingly being incorporated into government legislation [11] (**Table 1**).

2.4 National park

A national park is uniquely acknowledged for its conservation purposes. Especially, it is a reserve of natural, semi-natural, or developed land that a sovereign state declares or owns. Despite many efforts for individual nations to designate their own national parks differently, there is a common idea: the conservation of 'wild nature' for posterity and as a symbol of national pride [12]. An international organisation, the International Union for Conservation of Nature (IUCN) and its World Commission on Protected Areas (WCPA), has defined "National Park" as its Category II type of protected areas (see above) [11].

Category Ia: Strict Nature Reserve Category Ib: Wilderness Area	Category Ia: Strict nature reserve Protected areas that are strictly set aside to protect biodiversity and also possibly geological/geomorphological features, where human visitation, use and impacts are strictly controlled and limited to ensure protection of the conservation values. Such protected areas can serve as indispensable reference areas for scientific research and monitoring. Category Ib: Wilderness area Protected areas that are usually large unmodified or slightly modified areas, retaining their natural character and influence, without permanent or significant human habitation, which are protected and managed so as to preserve their natural condition.
Category II: National Park	Large natural or near natural areas set aside to protect large-scale ecological processes, along with the complement of species and ecosystems characteristic of the area, which also provide a foundation for environmentally and culturally compatible spiritual, scientific, educational, recreational and visitor opportunities.
Category III: Natural Monument or Feature	Protected areas set aside to protect a specific natural monument, which can be a landform, sea mount, submarine cavern, geological feature such as a cave or even a living feature such as an ancient grove. They are generally quite small protected areas and often have high visitor value.
Category IV: Habitat/Species Management Area	Protected areas aiming to protect particular species or habitats and management reflects this priority. Many category IV protected areas will need regular, active interventions to address the requirements of particular species or to maintain habitats, but this is not a requirement of the category.
Category V: Protected Landscape/Seascape	A protected area where the interaction of people and nature over time has produced an area of distinct character with significant ecological, biological, cultural and scenic value: and where safeguarding the integrity of this interaction is vital to protecting and sustaining the area and its associated nature conservation and other values.
Category VI: Protected Area with Sustainable Use of Natural Resources	Protected areas that conserve ecosystems and habitats, together with associated cultural values and traditional natural resource management systems. They are generally large, with most of the area in a natural condition, where a proportion is under sustainable natural resource management and where low-level non-industrial use of natural resources compatible with nature conservation is seen as one of the main aims of the area.

Source: https://www.iucn.org/theme/protected-areas/about/protected-areas-categories/category-ia-strict-nature-reserve.

Table 1.
IUCN protected area categories system.

To establish priorities for the protection of national parks, the IUCN established the primary objectives such as "to protect natural biodiversity along with its underlying ecological structure and supporting environmental processes, and to promote education and recreation". The interrelated objectives outlined are "to manage the area in order to perpetuate, in as natural a state as possible, representative examples of physiographic regions, biotic communities, genetic resources and unimpaired natural processes; to maintain viable and ecologically functional populations and assemblages of native species at densities sufficient to conserve ecosystem integrity and resilience in the long term; to contribute in particular to conservation of wide-ranging species, regional ecological processes and migration routes; to manage visitor use for inspirational, educational, cultural and recreational purposes at a level which will not cause significant biological or ecological degradation to the natural resources; to take into account the needs of indigenous people and local communities, including subsistence resource use, in so far as these will not adversely affect the primary management objective; and, to contribute to local economies through tourism" [11].

The IUCN World Parks Congress (WPC) is the world's premier global forum for setting the agenda for protected areas and takes place approximately once every 10 years. Since its first staging in 1962 in Seattle, USA, IUCN World Parks Congresses, previous IUCN Congresses have had a significant impact in assisting national governments to create new protected areas and directing more resources towards biodiversity conservation [13] (**Table 2**).

1	1962—Seattle, USA	The First World Conference on National Parks (Seattle, USA, 30 June–7 July 1962) aimed to establish a more efficient international understanding of national parks and to encourage further development of the national park movement worldwide. It set definitions and standards for representative systems of protected areas leading to the elaboration of the UN List of National Parks and Equivalent Reserves, later renamed UN List of Protected Areas.
2	1972—Yellowstone/Grand Teton National Park, USA	The Second World Conference on National Parks (Yellowstone, USA, 18–27 September 1972) focused on the effects of tourism on PAs; park planning and management; and social, scientific and environmental problems within national parks in wet tropical, arid and mountain regions. It also contributed to the genesis of the UNESCO World Heritage Convention and the Ramsar Convention on Wetlands of International Importance.
3	1982—Bali, Indonesia	The World National Parks Congress (Bali, Indonesia, 11–22 October 1982) focused on the role of PAs in sustaining society, and recognized 10 major areas of concern, including the inadequacy of the existing global network of terrestrial PAs and the need for: more marine, coastal and freshwater PAs; improved ecological and managerial quality of existing PAs; a system of consistent PA categories to balance conservation and development needs; and links with sustainable development.
4	1992—Caracas, Venezuela	The IVth World Congress on National Parks and Protected Areas: "Parks for Life" (Caracas, Venezuela, 10–21 February 1992) emphasized the relationship between people and PAs, and the need for, inter alia, the identification of sites of importance for biodiversity conservation, and a regional approach to land management. The Caracas Action Plan synthesised the strategic actions for PAs over the decade 1992–2002 and provided a global framework for collective action. The Plan aimed to extend the PA network to cover at least 10% of each major biome by 2000.
5	2003—Durban, South Africa	The Vth World Parks Congress (Durban, South Africa, 8–17 September 2003) held under the patronage of Nelson Mandela and Her Majesty Queen Noor of Jordan, the IUCN World Parks Congress 2003 helped develop a new paradigm for protected areas, defining and advancing the roles of governance, sustainable finance, capacity development, social equity and benefit sharing, leading to the Durban Action Plan and Durban Accord, both of which informed the Convention on Biological Diversity's successful Programme of Work on Protected Areas. Other outcomes included: the UN List and State of the World's Protected Areas, a global report on the world's PAs; a Protected Areas Learning Network (PALNet), a web-based knowledge management tool for PA managers and stakeholders; deliverables on Africa's PAs, including a recommendation on regional PAs and the Durban Consensus on African Protected Areas for the New Millennium; and a Handbook on Managing Protected Areas in the 21st Century, collating case studies, models and lessons learned during the Congress to constitute the "User Manual" for the Durban Accord.
6	2014—Sydney, Australia	The IUCN World Parks Congress 2014 (Sydney, Australia, 12–19 November 2014) focused on "Parks, people, planet: inspiring solutions." During the 8 days of plenary and workshop sessions, side events and field trips, participants addressed ways to: reach conservation goals; respond to climate change; improve health and well-being; support human life; reconcile different development challenges; enhance diversity and quality of governance; respect indigenous and traditional knowledge culture; and inspire a new generation to prioritize conservation.

Source: https://www.worldparkscongress.org/wpc/about/history

Table 2.
The IUCN World Parks Congress (WPC) has been convened six times at 10-year intervals since 1962.

Furthermore, various parties, that is policymakers, practitioners, CEOs, activists, and indigenous leaders, considered strategic issues related to PAs, conservation, and sustainable development in a series of seven moderated public debates, termed "World Leaders' Dialogues". The principal outcome document of the WPC, the Promise of Sydney, captured the main essence of the Congress as well as an ongoing online dialogue regarding potential solutions.

The 'ongoing' IUCN Red List of Threatened Species is acknowledged as the most high-profile contribution to conservation. The Red List is described by the IUCN as being "based on an objective system for assessing the risk of extinction of a species based on past, present, and projected threats" [13]. The stated goal of the Red List is to "provide information and analyses on the status, trends and threats to species in order to inform and catalyse action for biodiversity conservation" [14].

Launched in 2014, the IUCN Green List of Protected and Conserved Areas is the first global standard of best practice for area-based conservation. It is a designed certification for protected and conserved areas—national parks, natural World Heritage sites, community conserved areas, nature reserves, and so on—that are effectively managed and fairly governed [15]. Green-listed sites are certified as being effectively managed and fairly governed, with a positive impact on people and nature. The IUCN Green List also helps measure and accelerate progress towards Aichi Target 11, a UN biodiversity target that aims for 17% of terrestrial and 10% of marine areas to be protected, effectively conserved, and fairly managed by 2020. While the world is on track to meet the coverage aspect of Target 11, the 'effectiveness' aspect is still far from being reached. Since its launch, the number of countries committing to the IUCN Green List has grown fourfold—from 8 to 33 [15, 16]. Some 250 candidate sites have now volunteered to achieve its standard. The process of certification is voluntary and can take between 6 months and 5 years, during which time the sites work towards clear objectives and targets. For example, Van Long Nature Reserve, Vietnam, became a candidate in 2015. The certification process has helped secure commitment to expand the protected area into two neighbouring provinces.

Countries including Australia, Cote D'Ivoire, Kenya, Malaysia, Madagascar, as well as the US State of California have committed to nominating more sites to the IUCN Green List in the near future. California is prioritising its 124 marine protected areas, while Europe is studying how the standard could be implemented across its network of protected areas, Natura 2000. China, which counts six protected areas on the IUCN Green List, plans to nominate more sites (IUCN [4, 16]). IUCN just listed 15 new areas to the IUCN Green List of Protected and Conserved Areas, which now counts 40 sites in total [6].

By giving recognition to well-managed and well-governed protected and conserved areas, the IUCN Green List aims to increase the number of natural areas delivering long-lasting conservation results for people. Clearly, nature-protected and conserved areas are the foundation of biodiversity conservation. They safeguard nature and cultural resources, improve livelihoods, and drive sustainable development. The IUCN's ultimate scope of protected areas and conserved areas covers three key areas [6]:

- Achieving quality for successful and valuable protected areas
- Enhancing justice for fair, just, and inclusive protected areas
- Contributing protected area solutions to development challenges

The IUCN Programme 2017–2020 was of particular interest because it was approved by Member organisations at IUCN's World Conservation Congress in

September 2016 in Hawaii, USA. It was developed as a result of a nine-month consultation process across IUCN Members and Commissions. Accordingly, IUCN's work is guided by the Programme 2017–2020, which has three priority areas [6]: *first*, valuing and conserving nature enhances IUCN's heartland work on biodiversity conservation, emphasising both tangible and intangible values of nature; *second*, promoting and supporting effective and equitable governance of natural resources consolidate IUCN's work on people-nature relations, rights and responsibilities, and the political economy of nature; and *third*, deploying nature-based solutions to societal challenges expands IUCN's work on nature's contribution to tackling problems of sustainable development, particularly in climate change, food security, and social and economic development.

3. The threats to NPs and PAs

In particular, there are cases around the world where protected areas face serious environmental problems. The management of protected areas is a complicated process that often reflects the conflict in the relationship between society and protection, usually through the prohibition of particular activities as established in the existing protection status [17].

The loss of biodiversity can be attributed to extinction that refers to the cessation of the existence of a species or a group of species. It is a natural process that is when species are diminished because of environmental forces (habitat fragmentation, global change, natural disaster, and overexploitation of species for human use) or because of evolutionary changes in their members (genetic inbreeding, poor reproduction, and decline in population numbers). The moment of extinction is generally when the death of the last individual of a particular species occurs. In many cases, however, the capacity to breed and recover may have been lost before this point [8].

Deforestation has been seen around the world in decades, though tropical rainforests are particularly targeted. If current deforestation levels proceed, the world's rainforests may be globally extinct in as little as 100 years, according to National Geographic. According to GRID-Arendal—a United Nations Environment Program collaborating center, a frequent deforestation trend occurred in countries particularly Indonesia, Thailand, Brazil, the Democratic Republic of Congo, and other parts of Africa, as well as some parts of Eastern Europe. As a consequence, several sites within the protected area network have been frequently degraded. According to a study by the University of Maryland and the World Resource Institute [18], Indonesia is recorded the most deforested nation. Since the last century, the country has lost at least 39 million acres (15.79 million hectares) of forest land. Similarly, African national parks also faced with deforestation problems and continuous reduction of forest lands mainly due to political actions as well as the application of measures applied by those responsible for the management of these parks [19]. The terrible damage in protected areas, despite the strict protection status [20], is linked with social factors such as the increased logging activity, which means more revenue for rural counties, where logging mills are often located, and more jobs in these areas. Many rural communities have experienced economic decline since environmental concerns decreased logging on federally protected lands [21]. In Costa Rica, particularly around the Osa Peninsula, the greater forest cover loss in areas with a high density of threatened, endemic, and new-to-science species poses a significant question regarding the success of future conservation initiatives outside protected areas [22]. The UK, similar to many of the countries such as the USA, Caribbean,

Latin America, Australia, and the Mediterranean, suffers from coastal flooding and erosion, as a consequence of intensive industrial and agricultural activities, as well as by urbanisation and tourism [23, 24]. As [25] state, intensive logging constitutes a factor that provides further evidence for anthropogenic disruption of the structure and functioning of rainforest ecosystems.

In the case of the National Park System, fires last year burned through parts of Glacier, Yosemite, Crater Lake, Grand Canyon, and Yellowstone National Parks, as well as Whiskeytown and Santa Monica Mountains National Recreation Areas [26]. In Laguna del Tigre National Park in Guatemala, about two thirds of the total land of the park was destroyed by forest fires, as a consequence of phenomena associated with considerable immigration of population to the area as well as with deficiencies affiliated with organisation issues of the administration bodies [27]. On the other hand, according to [28], in Arikok National Park, Aruba, the herbivory causes vegetation cover to decrease, which can lead bare soils to become vulnerable for wind and water erosion problems, which can be attributed to trampling and grazing of livestock.

In Cardamom National Park protected area, Wildlife Alliance Rangers raided two illegal settlements and confiscated two homemade guns, one chainsaw, electric fishing gear, and wild animals' parts. After examining the impact of hunting on the distribution of biodiversity in the protected area, [29] concluded that mainly the settlements and their occupants had been conducting forest crimes such as large-scale logging and poaching since 2016. This means that the Wildlife Alliance's Forest Protection Program will need to counter these threats by stopping poachers and loggers on daily basis [30]. Also, the problem of illegal hunting created debates in relation to effective management and conservation of biodiversity with regard to the protected areas of Africa [31].

According to a new study published recently, the authors have warned that climate change has adversely and uniquely affected many of the 417 national parks spread across the United States and its territories [32]. A study published in Nature, a leading scientific journal, provides data that suggest that climate change–related phenomena have killed 150,000 people annually for the past 30 years and that numbers will increase [33].

The long-lasting state of economic and political crisis, economic sanctions combined with consequences of war, and the NATO air strikes in 1999 resulted in an enormous destruction of infrastructure and high foreign debts in some countries in Europe [1]. Due to this reason, protected areas in Serbia, for instance, suffer from inadequate funding. In correspondence with the existing legal regulation, some portions of the national budget must be allocated for protected areas, besides fund for environmental protection, taxes for use of natural resources, income from their own activities, projects, and other donations.

In the case of the protected area in Serbia, [34] have pointed out that the financial allocation of the Ministry of Environment and Spatial Planning (MESP) covers limited part of the operational costs of the protected area and its portion is dwindling. Therefore, most of the protected areas (PA) in Serbia are faced with financial constraint and not sustainable [1]. In addition, the capacity development processes and projects are oriented towards setting up standards of competences enabling professional staff and empowering community actors to meet high demands of modern protected area management. Moreover, according to [35], the preservation of the KwaZulu-Natal PA residents with high illiteracy, poverty, and unemployment like those in Kwadlangezwa are willing to participate in ecotourism businesses, as this could be a way of empowering them economically. Arguably, willingness to participate in ecotourism development and planning could be linked to the higher level of literacy in Empangeni.

Furthermore, the National Chambal Sanctuary (NCS), otherwise known as the National Chambal Gharial Wildlife Sanctuary, is a PA in northern India—which flows in three states of India: Uttar Pradesh (UP), Madhya Pradesh (MP), and Rajasthan and covers in and around the Chambal River—and is facing various issues related to social, economical, environmental, and institutional factors [36].

In addition, comparing their first study results, where tourism development, biodiversity protection, and climate change were mostly regarded separately, a shift towards a more integrated development, including climate change as one of the issues to be solved, was observed [37]. Moreover, an integrated solutions found in participatory scenario approaches can be an "eye-opener" for climate change adaptation and act as a tool to unblock the elements of the motivation chain, and thus end up by generating action.

Although incentives for the protection of the environment are without a doubt important and decisive for the future of the planet, it is acknowledged that within a broader frame of governance, landscape protection and natural and cultural heritage conservation are linked to training opportunities and skills enhancement, physical and intellectual access, community participation, and the engagement of local residents and visitors [38]. Decision-making processes will always include to some extent conflicting interests that must be balanced in order for problems to be solved [17].

It is common to find the fact that no specific planning organisations and routines exist for climate change adaptation, integrated strategic planning combining biodiversity, and PA management. If used partially, for example, in regional planning procedures, these approaches can be carried out at a conceptual level. In consequence, this can lead to conflicts of interests between the different forms of human activity, thereby leading to less impactful outcomes. In Greece, with the absence of executive authority, the management bodies of protected areas are ineffective in performing their administrative and management duties. Such executive authority is given to other related bodies and control mechanisms [39], a fact that in most cases creates confusion regarding the responsibilities of each body especially during the implementation of protection measures. In consequence, this has led to increasing conflict among the multiple authorities and breaches of the legal requirements for environmental protection that apply in the area or people's own ability to take action is very limited.

4. Some global initiatives on the management of protected areas

The management of protected areas is a complicated process that often reflects the conflict in the relationship between society and protection, usually through the prohibition of particular activities as established in the existing protection status [17]. The countries have become more directly involved in the conservation of biodiversity such as the prohibition of trade in endangered species, the establishment of protected areas, and the crafting and enforcement of laws to regulate land use. But, to conserve biodiversity, there has also been increased reliance on economic instruments [8]. The following are some of the relevant global initiatives:

4.1 The United Nations Convention on Biological Diversity (CBD)

During the Earth Summit in Rio de Janeiro in 1992, the United Nations Convention on Biological Diversity (CBD) was committed to stem the rapid worldwide loss of biodiversity and to provide a legal framework for its conservation. The three indicators for success include to promote the conservation of biological diversity, the sustainable use of its components, and the fair equitable sharing of the benefits arising from the use of genetic resources [15].

4.2 Natural World Heritage

Natural World Heritage sites are recognised as the planet's most important protected areas, providing preparedness and resilience to millions of people worldwide [3]. And yet, they are under the specificities of different risks from mining, poaching, climate change, infrastructure development, and other threats. Natural World Heritage sites include iconic places such as the Great Barrier Reef, Yellowstone, the Galápagos Islands, and Kilimanjaro and are recognised as the world's most significant protected areas. IUCN's evaluations of nominations are part of a rigorous process where a wide range of information is reviewed. IUCN's World Heritage Panel is an essential part of this process, ensuring the highest quality of independent advice.

4.3 World commission on protected areas

The Commission is recognised as the premier network of 2500 experts from 140 countries that mobilises action in science, conservation, policy, and engagement to support well-managed and connected parks and other protected areas. The Commission develops knowledge-based policy, advice, and guidance on the full suite of issues surrounding protected areas through the establishment of specialist groups and task forces, as well as global protected area standards and best practice guidelines (see more details at [16]).

4.4 Related solutions for protected areas

In facing the most pressing governance challenges of natural hazard, protected areas offer vital solutions to some issues including species extinction, climate change, and poverty. Acknowledging this, IUCN proposed 'Solutions' as one of the international scale joint frameworks and cooperation strategies for its global work on protected areas. For more details, see reference [3, 16].

4.4.1 World Conservation Congress

With the goal of conserving the environment and harnessing the solutions nature offers to global challenges, the IUCN World Conservation Congress becomes the forum for several thousand leaders and decision-makers from government, civil society, indigenous peoples, business, and academia [5].

The Congress contributes to a subset of managing natural environment for human, social, and economic development, but this cannot be achieved by conservationists alone. The IUCN Congress should not be viewed as an isolated means to address differences and work together to create good environmental governance, but rather to engage in all parts of society to share both the complex and multifaceted part of conservation. The IUCN World Conservation Congress convenes every 4 years to set priorities and agree on the Union's work programme. IUCN congresses have produced several key international environmental agreements including the Convention on Biological Diversity (CBD), the Convention on International Trade in Endangered Species (CITES), the World Heritage Convention, and the Ramsar Convention on Wetlands [5].

4.4.2 Oceania

Oceania is geographically one of IUCN's largest regional programmes, covering over 100 million square kilometres of the Pacific Ocean.

4.4.3 Biodiversity and Protected Areas Management (BIOPAMA) Programme

The Biodiversity and Protected Areas Management (BIOPAMA) Programme is a €60 million initiative of the European Union (EU) and the African, Caribbean and Pacific (ACP) Group of States to improve the long-term conservation and sustainable use of natural resources through better use and monitoring of information and capacity development on management and governance. It is implemented through IUCN (International Union for Conservation of Nature) and the Joint Research Centre of the European Commission (EC-JRC). The programme focuses on the 79 ACP countries, which cover a huge diversity of ecological, social, and cultural systems. These regions contain a significant proportion of our planet's biodiversity.

4.4.4 PANORAMA—solutions for a healthy planet

PANORAMA is a partnership promoting examples of inspiring, replicable solutions across a range of conservation and development topics, to enable cross-sectoral learning and upscaling of successes.

PANORAMA, through its "protected areas solutions" thematic community, profiles specific case studies of such solutions, their success factors, and lessons learnt. It also places protected areas in a larger thematic context, through other, inter-linked PANORAMA thematic communities, profiling solutions across multiple themes.

IUCN has acknowledged the role that protected areas can play, for example, in climate change adaptation and mitigation, disaster risk reduction, ensuring sustainable local livelihoods, and addressing desertification. This results in well-planned and justly governed areas, when the costs and benefits of protected area systems are shared equitably, as well as mainstreamed into all economic sectors. Consequently, protected areas contribute to national and local economies. In particular, these approaches are the foundation for sustainable and resilient livelihoods for many communities.

4.4.5 MOOCs and protected areas capacity building

Massive open online courses (MOOCs) are now an undeniable part of the education landscape and a revolutionary opportunity for everyone across the world with an Internet connection to access free courses and to receive a certificate or academic credits. The IUCN Papaco MOOCs on 'protected areas management' and 'ecological monitoring' have cumulated a number of 12,700 registrations so far from more than 120 countries. Two more MOOCs have just started, on 'law enforcement in protected areas' and 'species conservation' in African protected areas.

4.4.6 CapeNature

The Protected Area Solutions project relies on the IUCN Green List of Protected and Conserved Areas (GLPCA) to deliver on the "quality" elements of Aichi Target 11. The project also develops a portfolio of case studies to communicate successful PA biodiversity and climate change outcomes through the PANORAMA platform.

Closely related to the mission of PANORAMA, the WCPA Natural Solutions specialist group promotes and develops the potential for protected areas to deliver a range of ecosystem services, with a particular focus on food and water security and disaster risk reduction.

4.4.7 Additional solutions for protected areas

Some of the additional solutions for protected areas include the following:

- *Clean Water:* One third of the 105 world's largest cities derive drinking water from forested protected areas. Another 10% obtain water from sources that originate in 'protected' watersheds. Some natural forests (especially tropical montane cloud forests) and wetland areas increase total water flow.

- *Food Security*: Protected areas conserve populations of wild species, pollinators, and crop wild relatives safeguarding species and populations vital for food security and agriculture.

- *Health:* Climate change and lack of clean water are expected to lead to increased health risks and epidemics. Protected areas contribute to health by protecting intact ecosystems and supplies of medicinal plants and genetic material for pharmaceuticals. Many people (80% in Africa) rely on traditional medicines; 28% of plants are used medicinally; and 60% of medicinal plants are collected from the wild, including in protected areas.

- *Ecosystem stability*: Economic losses from natural disasters have increased 10-fold over the last 50 years. Protected areas can play a role in helping reduce the occurrence and impacts of natural disasters, such as floods, landslides, tsunamis, typhoons and storms, fire and drought, and desertification. For example, in Argentina, flood protection programmes have integrated conservation of natural habitats to reduce vulnerability and disaster risks, to complement infrastructure and early-warning investments.

- *Distinguishing features*: Category II areas are typically large and conserve a functioning "ecosystem", although to be able to achieve this, the protected area may need to be complemented by sympathetic management in surrounding areas.

    - The area should contain representative examples of major natural regions, and biological and environmental features or scenery, where native plant and animal species, habitats, and geodiversity sites are of special spiritual, scientific, educational, recreational, or tourist significance.

    - The composition, structure, and function of biodiversity should be to a great degree in a "natural" state or have the potential to be restored to such a state, with relatively low risk of successful invasions by non-native species.

- *Role in the landscape/seascape*: Category II provides large-scale conservation opportunities where natural ecological processes can continue in perpetuity, allowing space for continuing evolution. They are often key stepping-stones for designing and developing large-scale biological corridors or other connectivity conservation initiatives required for those species (wide-ranging and/or migratory) that cannot be conserved entirely within a single protected area.

5. Summary and conclusion

Biodiversity needs to be conserved, protected, and used sustainably. The benefits that biodiversity provides to society require that effective measures should be managed appropriately and, when indicated, coordinated with ongoing scientific projects. This is to ensure that nature is protected and used sustainably. Today, however, biodiversity is facing a variety of environmental impacts that are directly or indirectly attributable to a variety of environmental impacts due to the concentration of a wide range of human activities and rapid growth in human population. In many parts of the world, however, the loss of biodiversity has been identified as a serious environmental problem that can significantly undermine the prospects for sustainable development and the evolution of ecosystem and reduces their resilience. Our analysis shows that the most significant elements of sustainability in protected area governance that should receive highest priority are integrated approach to natural, cultural, and economic aspects; flexibility in responding to constant changes and pro-activity in finding innovative solutions; involving local community in planning, decision-making, and providing economic benefits with no harm to nature; respecting traditional knowledge as the basis for building up new learning approaches and programmes; and connectivity to other protected areas, common issues, and international developments [1]. It is worth noting, therefore, that the national parks and other types of protected areas continue to form the basis of the full range of species and natural ecosystems that must be better designed, protected, and managed despite impending weaknesses regarding guarding procedures, supervision, and implementation of protection measures in all these areas.

In addition to IUCN's close collaboration with UNESCO and WWF, this fact shows that there is a need for a better organisation and co-operation among the administrative and management bodies of a country that will require the cooperation of international agencies, non-governmental organisations, researchers, as well as local communities in the areas of conservation or any other protected area authorities. This approach, nonetheless, may be modified to specific contexts and goals, and applied with this purpose to other similar social environments, especially in countries in transition within the region. With this kind of intra- and inter-kind of relationships between various stakeholders, all parties will strengthen their positive attitude towards preserving and also lend greater support in the wider scope of PAs and NPs, since they think that essentially it has positive influence on the life of the inhabitants and on the natural environment.

Acknowledgements

We would like to express our gratitude to IntechOpen for their support throughout the stage of preparing this book. Thanks also go to the Faculty of Administrative Science & Policy Studies (FSPPP), Research and Management Centre UiTM, Centre for Bio-Diversity & Sustainable Development, and UiTM Johor Malaysia for facilities and resources granted to us in preparing this book. We appreciate all suggestions, observations, and input from all contributors to this volume, as well as the unidentified reviewers chosen by IntechOpen in evaluating the manuscripts. May this book be helpful and useful to them and enrich everyone interested in the field of national parks and protected areas towards future sustainability.

Author details

Ahmad Naqiyuddin Bakar[1,2,3,4]

1 Faculty of Administrative Science and Policy Studies, UiTM, Shah Alam, Selangor, Malaysia

2 Malaysia Institute of Transport (MITRANS), UiTM, Shah Alam, Selangor, Malaysia

3 Centre for Biodiversity and Sustainable Development, UiTM, Puncak Alam, Selangor, Malaysia

4 UiTM Johor Branch, Segamat Campus, Malaysia

*Address all correspondence to: ansbakar@uitm.edu.my

References

[1] Orlović-Lovren V. The role of education in protected area sustainable governance. Management of Environmental Quality. 2011;**22**(1):48-58

[2] United Nations—UN. The Sustainable Development Agenda [Internet]. 2019. Available from: https://www.un.org/sustainabledevelopment/development-agenda/ [Accessed: 23-02-2019]

[3] IUCN—International Union for Conservation of Nature. IUCN Director General's statement for World Environment Day and World Oceans Day 2018 [Internet]. 2018c. Available from: https://www.iucn.org/news/secretariat/201806/iucn-director-general%E2%80%99s-statement-world-environment-day-and-world-oceans-day-2018 [Accessed: 13-02-2019]

[4] IUCN—International Union for Conservation of Nature. Report shows 15% of terrestrial and 7% of marine areas now covered by protected areas [Internet]. 2018b. Available from: https://www.iucn.org/news/secretariat/201811/report-shows-15-terrestrial-and-7-marine-areas-now-covered-protected-areas [Accessed: 13-02-2019]

[5] IUCN—International Union for Conservation of Nature. A brief history [Internet]. 2019. Available from: https://www.iucn.org/about/iucn-brief-history [Accessed: 27-02-2019]

[6] IUCN. IUCN Protected Area Categories System [Internet]. 2018. Available from: https://www.iucn.org/theme/protected-areas/about/protected-areas-categories/category-ia-strict-nature-reserve [Accessed: 13-02-2019]

[7] Bongaarts J. United nations department of economic and social affairs, population division world mortality report 2005. Population and Development Review. 2006;**32**(3):594-596

[8] Ayoo C. Economic instruments and the conservation of biodiversity. Management of Environmental Quality. 2008;**19**(5):550-564

[9] Myers N. The Sinking Ark: A New Look at the Problem of Disappearing Species. Oxford, England, UK: Pergamon Press; 1980. xiii + 305 pp., 9 tables

[10] Barnosky A, Matzke N, Tomiya S, Wogan G, Swartz B, Quental T, et al. Has the Earth's sixth mass extinction already arrived? Nature. 2011;**471**:51-57

[11] IUCN—International Union for Conservation of Nature. Nature's contribution to sustainable development takes centre stage in Geneva [Internet]. 2018. Available from: https://www.iucn.org/news/secretariat/201810/natures-contribution-sustainable-development-takes-centre-stage-geneva [Accessed: 13-02-2019]

[12] EUROPARC Federation, editor. 100 years of National Parks in Europe: A shared inheritance, a common future. In: Report from the EUROPARC Federation Conference; 8-13 September 2009, Strömstad, Sweden Oekom Verlag, München. 2009. pp. 1-24

[13] IUCN. What is the IUCN World Parks Congress? [Internet]. 2019. Available from: https://www.worldparkscongress.org/wpc/about/history [Accessed: 27-02-2019]

[14] IUCN—International Union for Conservation of Nature. IUCN Green List of Protected and Conserved Areas [Internet]. 2019b. Available from: https://www.iucn.org/theme/protected-areas/our-work/iucn-green-list [Accessed: 27-02-2019]

[15] IUCN—IUCN reviews nine new World Heritage nominations for 2019 [Internet]. 2019. Available from: https://www.iucn.org/news/world-heritage/201812/iucn-reviews-nine-new-world-heritage-nominations-2019 [Accessed: 27-02-2019]

[16] IUCN—International Union for Conservation of Nature. IUCN lists 15 new sites as world's best protected areas [Internet]. 2018b. Available from: https://www.iucn.org/news/protected-areas/201811/iucn-lists-15-new-sites-worlds-best-protected-areas [Accessed: 13-02-2019]

[17] Veronika A, Stilianos T, Georgios T, Evangelos M. Environmental problems in protected areas: Stakeholders' views with regard to two neighboring National Parks in Greece. Management of Environmental Quality. 2014;**25**(6):723-737

[18] Bradford A. Deforestation: Facts, Causes & Effects. [Internet]. 2018. Available from: https://www.livescience.com/27692-deforestation.html [Accessed: 27-02-2019]

[19] Petursson JG, Vedeld P, Sassen M. An institutional analysis of deforestation processes in protected areas: The case of the transboundary Mt. Elgon, Uganda and Kenya. Forest Policy and Economics. 2013;**26**:22-33

[20] Oestreicher JS, Benessaiah K, Ruiz-Jaen MC, Sloan S, Turner K, Pelletier J, et al. Avoiding deforestation in Panamanian protected areas: An analysis of protection effectiveness and implications for reducing emissions from deforestation and forest degradation. Global Environmental Change. 2009;**19**(2):279-291

[21] Palmisano J. Logging in National Parks and Forests: A Contentious Debate [Internet]. 2014. Available from: https://lawstreetmedia.com/issues/energy-and-environment/should-logging-be-encouraged-in-national-parks-and-forests-under-hr-1526/ [Accessed: 27-02-2019]

[22] Sanczhez-Azofeifa GA, Rivard B, Calvo-Alvarado J, Moorthy I. Dynamics of tropical deforestation around National Parks: Remote sensing of Forest change on the Osa peninsula of Costa Rica. Mountain Research and Development. 2002;**22**:352-358. DOI: 10.1659/0276-4741(2002)022[0352,DOTDAN]2.0.CO;2

[23] William A, Guillermo Rangel-Buitragoc N, Pranzini E, Anfuso, G. The management of coastal erosion, Ocean & Coastal Management [Internet]. 2017. Available from: http://dx.doi.org/10.1016/j.ocecoaman.2017.03.022 [Accessed: 23-02-2019]

[24] Skanavis C, Sarri E. The role of environmental education as a tool for environmental management in Cyprus: Strategies and activities. Environmental Management and Health. 2002;**13**(5):529-544

[25] Woodcock P, Edwards DP, Newton RJ, Vun Khen C, Bottrell SH, Hamer KC. Impacts of intensive logging on the trophic organisation of ant communities in a biodiversity hotspot. PLoS One. 2013;**8**(4):e60756. DOI: 10.1371/journal.pone.0060756

[26] Repanshek K. New Executive Order Could Open Forests In National Parks To Logging [Internet]. 2019. Available from: https://www.nationalparkstraveler.org/2019/01/new-executive-order-could-open-forests-national-parks-logging [Accessed: 27-02-2019]

[27] Monzón-Alvarado C, Cortina-Villar S, Schmook B, Flamenco-Sandoval A, Christman Z, Arriola L. Land-use decision-making after large-scale forest fires: Analyzing fires as a driver of deforestation in

Laguna del Tigre National Park, Guatemala. Applied Geography. 2012;**35**(1-2):43-52

[28] Veerbeek B. Goats or no Goats: The Influence of Goats on Soil Erosion and Vegetation in Arikok National Park, Aruba [Thesis]. The Netherlands: Wageningen University; 2016

[29] Koh Kong, Wildlife Alliance. Poaching and logging crackdown inside Cardamom National Park protected area. [Internet]. 2018. Available from: https://www.wildlifealliance.org/poaching-and-logging-crackdown-in-koh-kong/ [Accessed: 27-02-2019]

[30] Rainforest Protection [Internet]. 2019. Available from: https://www.wildlifealliance.org/ [Accessed: 23-02-2019]

[31] Johannesen AB. Protected areas, wildlife conservation, and local welfare. Ecological Economics. 2007;**62**(1):126-135

[32] Gonzalez P, Neilson RP, Lenihan JM, Drapek RJ. Global patterns in the vulnerability of ecosystems to vegetation shifts due to climate change. Global Ecology and Biogeography. 2010;**19**(6):755-768

[33] Patz Jonathan A, Campbell-Lendrum D, Holloway T, Foley A. J. Impact of regional climate change on human health. Nature. 2005;**438**:310-317

[34] Laktić T, Pezdevšek Malovrh Š. Stakeholder participation in Natura 2000 management program: Case study of Slovenia. Forests. 2018;**9**(10):599

[35] Adeleke BO. Assessment of residents' attitude towards ecotourism in KwaZulu-Natal protected areas. International Journal of Culture, Tourism and Hospitality Research. 2015;**9**(3):316-328

[36] Yadav N, Sahu NC, Sahoo D, Yadav DK. Analysis of barriers to sustainable tourism management in a protected area: A case from India. Benchmarking: An International Journal. 2018;**25**(6):1956-1976

[37] Lupp G, Heuchele L, Renner C, Syrbe R, Konold W, Siegrist D. Motivations and attitudes to (not) take action for climate change adaptation in protected areas. International Journal of Climate Change Strategies and Management. 2016;**8**(3):356-374

[38] Clarke R. Beyond landscape designation: Innovative funding, delivery and governance and the UK protected area system. Management of Environmental Quality. 2015;**26**(2):172-194

[39] Vokou D, Vareli K, Zarali E, Karamanoli K, Constantinidou HIA, Monokrousos N, et al. Exploring biodiversity in the bacterial community of the Mediterranean phyllosphere and its relationship with airborne bacteria. Microbial Ecology. 2012;**64**(3):714-724

Chapter 2

The Effect of Forest Institution Connectedness, Incentive Participation Program, and Social Capital on Public Participation and Welfare as Mediators of Forest Management in Baluran National Park

Adil Siswanto and Djumilah Hadwidjojo

Abstract

The effect of forest institution connectedness, incentive participation program, and social capital on public participation and welfare as mediators of forest management were conducted in the forest conservation area of Baluran National Park, East Java, Indonesia. The problem facing this area is that the five buffer villages of Wonorejo, Sumber Waru, Sumber Anyar, Bajul Mati, and Watu Kebo exploit the forest's resources for their own economic reasons. The purposes of this research are to analyze and explain: (1) the effect of the forest institutions connectedness on public participation; (2) how welfare mediates forest institutions connectedness to public participation; (3) the effect of incentive participation programs on public participation; (4) how welfare mediates the effect of incentive participation programs on public participation; (5) the social effect capital on public participation; (6) how welfare mediates social capital's effect toward public participation; and (7) welfare's effect on public participation. The survey method and questionnaires were used for a proportional random sampling of 170 respondents. They are 120 households that were members of the forestry community training center and 50 respondents from the staff of Baluran National Park. Validity and reliability testing of instruments and hypothesis were performed using WarpPLS 5.0 software. The results show: (1) forest institutions connectedness to public participation does not contribute positive significant effect; (2) the effect of forest institution that is related to public participation has been fully mediated by welfare; (3) incentive participation programs have a positive significant effect on public participation; (4) welfare partially mediates the effect of incentive participation programs on public participation; (5) social capital has a positive significant effect on public participation; (6) welfare mediates social capital's effect on public participation; and (7) welfare has a positive significant effect toward optimizing public participation in forest conservation management in the Baluran National Park. Practical implications of this research are: (1) the contribution of nontimber forest products as a proportion of families' income is between 12.99% and 28.46%; and (2) based on the classification of public participation especially in four programs (participation in planning program, implementation, benefit-sharing, and evaluation and monitoring) that are low level at 47.1%, middle level at 33.5%, and high level at 19.4%.

Keywords: incentive participation program, social capital, welfare, public participation, social forestry management

1. Introduction

A national park, as a forest conservation area, has a variety of flora and fauna which can be relied upon to ensure the human survival for now and future [1]. The majority of these parks have now faced threats and interferences such as encroachment, and illegal cultivation continues to increase over time [2]. Threats and disturbances in these areas are caused by various factors, namely (1) the institutions role in forest conservation management and local population participation level of the are still not optimum (especially in the case of those living around the forest); (2) the lack awareness about the conservation area is still very low among local people; (3) the education level of local people is low; and (4) there is a lack of agricultural land [1, 3].

The forest destruction in Baluran National Park includes: (1) forest fires in 2014, with 132 fires covering an area of around 2005.90 ha. Rather than natural factors, the main causes of forest fires are local people not acting responsibly, a lack of security personnel guarding the forest, and weak law enforcement. Forest fires impact heavily on the flora and fauna. (2) Clearing activities as a result of 400 ha being devoted to agricultural plants business. (3) Timber theft (as well as theft of firewood, fruit tart, hazelnut, *gebang* trees, ornamental fish and over grassing) especially in the Labuhan Merak resort. (4) Cattle grazing is a problem that is quite prominent, especially in the areas of Karangtekok, Labuhan Merak, and Balanan with about 3450 ha. Cattle grazing (cows and goats) is widespread, with an average of 1447 head of cattle per day. As a result of this illegal grazing, the soil becomes solid, which is harmful to plants and vegetation that could potentially be survival disruption of the park, as well as deer, antelope, and bison (the unique wildlife of Baluran National Park). (5) Local transmigration settlements since 1976, covering an area of 57 ha in Pandean area of Wonorejo village. (6) Illegal encroachment and the tilling of the soil. (7) Hunting of wildlife by people with firearms, snares, poison, and sap that often occurs during the dry season. Various factors affect the behavior and movement patterns of animals, including a limited source of drinking water for animals, especially in the dry season. Based on the above phenomena, this paper focuses on the damaged forest in the Baluran National Park, caused by the poor level of public participation [4–6].

There are some previous studies which discuss public participation in the forestry management program. These include: (1) studies which explained the factors affecting public participation in forestry management, because of the role of forest institution connectedness by Baynes et al. [7], Muro and Namusonge [8] and Lise [9], (2) the quality of forest institutions as good governance will be able to create conditions of security, belief, trust, and economic welfare by Hans-Jurgen [10] and Akib et al. [11], (3) the effects of incentive participation program for social forestry management in increasing public participation by Adhikari et al. [12], Djamhuri [13], and Kaseya and Kihonge [14], (4) incentive participation program effects in relation to welfare Rahut et al. [15], William and Ayuk [16], Das and Sarker [17], (5) the social capital effect on public participation by Sara et al. [18] and Sharpe

[19], (6) social capital effect on public welfare by Grootaet [20] and Narayan and Pritchett [21], (8) how welfare effect on public participation in development programs Rahut et al. [15] and Akamani and Hall [22].

The issue of differences in measurement of incentive participation program is a gap in this research. Adhikari et al. [12] measured incentive participation program with six indicators, they are: (1) access to forests and availability of forest products, (2) financial support to supplement household income, (3) social security and cohesion through local institution building, (4) investment in local community infrastructure and development, (5) well-defined & enforced property rights over forest resources assigned to the users, and (6) payment for environmental services. The research's purposes of Adhikari et al. [12] are: (1) to determine the relationships between different incentive participation program and the level of public participation of user group members; (2) to explore how households might respond to any changes in the incentive participation program, in terms of their decision to participate in common property resource governance; and (3) to propose/recommend how organizational incentive participation program can be better integrated in order to induce more effective public participation of users in the governance and management of property resources. The indicators of public participation were measured based on (1) membership length; (2) representation on the executive committee; (3) level of public participation in meetings, (4) in decision-making, and (5) in implementation; and (6) overall benefits.

While Djamhuri [13] measured incentive participation program with seven indicators, they are: (1) forest village population); (2) villages forests/WPH; (3) number of forest village community (LMDH) trustee board members; (4) percentage of Tumpang Sari Farmers on the LMDH trustee board; (5) tree coverage on foundation of the LMDH; (6) current tree coverage; (7) trustee board members attendance of routine meetings. The indicative numbers of LMDH trustee board members and percentage of Tumpang Sari Farmers on LMDH trustee board consist of: (1) formal education; (2) household annual income; (3) use of feed/fodder from state forest land; and (4) use of firewood from forest land. Djamhuri [13] said Tumpang Sari is an incentive participation program which is traditional in forest management. Government and society integration provides a better incentive participation program in the hope that the public will be will contribute in the state forest management.

Kaseya and Kihonge [14] measured incentive participation program with three indicators, they are: (1) civic education, (2) financial incentives both transport and lunch allowances, and (3) scheduling of forums/meetings. The study result was corroborated by the findings from the open interview which indicated that 62.5% of the respondents concurred that financial incentives are offered to participants. Measurements of incentive participation program in this research refers to [12], but its indicators are based on research object conditions.

The second gap of this research is the differences of social capital's measurements done by Grootaet [20] and Narayan and Pritchett [21]. Grootaet [20] measured social capital into six dimension of social capital, they are: (1) density of membership, (2) heterogeneity index, (3) meeting attendance, (4) decision making index, (5) membership dues, and (6) community orientation.

Narayan and Pritchett [21] measured social capital into six variables, they are: (1) heterogeneity members, (2) inclusiveness members, and (3) performances members. Social capital's indicators consist of: (1) membership, (2) characteristic of membership; (3) values and individual's **behaviors**.

Measurement of social capital in this study refers to Grootaet [20], who measures social capital as a factor in the reduction of poverty and increase in prosperity, but indicators of social capital of this research based on research object condition. Welfare provision would increase the role of public participation in development.

Increased public participation will reduce transaction costs and the cost of control, raise output and further improve the welfare of the community.

Rahut et al. [15] suggested increasing public participation in collaborative forest management (CFM) while adding welfare as a mediation variable which will affect social capital and also affect public participation.

The novelty of research are: (1) to examine the integrative model with purposes to insure weather forest institution connectedness, incentive participation program, social capital and public welfare still have positive significant effect on public participation; (2) to analyze the level of public participation based on participation in planning program, implementation, benefit-sharing; and evaluation and monitoring of the forest conservation management in Baluran National Park.

After all, the purposes of this study that were presented here will be to analyze and explain: (1) effect of the forest institutions' connectedness to public participation; (2) how welfare mediates forest institution connectedness to public participation; (3) the effect of incentive participation programs on public participation; (4) how welfare mediates the effects of incentive participation programs on public participation; (5) social capital's effect on public participation; (6) how welfare mediates social capital's effect on public participation; and (7) welfare's effect on public participation.

2. Material and method

Data were collected during July–December 2017 through interview, research questionnaires, and documentation. Interview was conducted to determine the respondents' answers to a questionnaire relating to the variables that have been used in this study.

The sampling method is proportional random sampling. The unit analysis is the heads-of-household who are members of the forestry community training center (120 people) and the staff of Baluran National Park with 50 people. All of them are 170 respondents in total (see **Table 1**). The construct validity of reflective indicators were tested based on convergent validity, discriminant validity, composite reliability [23]. Variable with formative indicators were tested based on the values of full collinearity variances inflations factor.

The method of data analysis used in this study is structural equation modeling using WarpPLS 5.0. This research is based on working with numbers, and the data are tangible, analyzed using statistics to test hypotheses or answer specific research questions and to make predictions that a particular variable affects other variables [24].

To test mediation roles the causal-step approach of Baron and Kenny was used. The best way to test for mediation effects is by counting the Variance Accounted For (VAF) value, which can determine the indirect effect relative to the total effect [25].

According to Baron and Kenny [26] the causal step approach has four mediation effects, they are: (1) nonmediation, if VAF value $< 20\%$; (2) partial mediation, if VAF value is around $20 \leq 80\%$; (3) full mediation, if VAF value $> 80\%$; and (4) suppressed mediation, if the direct effect sign changed after inclusion of the mediation variable.

3. Result and discussion

3.1 Validity test

The validation of reflective indicators was done through: (1) convergent validity; (2) discriminant validity; and (3) reliability test consists of (a) indicator reliability; and (b) consistency internal reliability both composite reliability and Cronbach alpha [23].

3.2 Convergent validity

Convergent validity testing is performed to identify the items of instrument indicators as indicators from a latent variable (see **Table 2**). The convergent validity test result shown that all of the outer loading values are more than 0.6 (>0.6). At last, it can be seen that this research has met the requirements of the convergent validity [23, 25].

No	Name of Village	Sample Size Needed
1	Wonorejo Village	62/170 x 120 = 43 people
2	Sumber Anyar Village	50/170 x 120 = 35 people
3	Sumber Waru Village	33/170 x 120 = 24 people
4	Watu Kebo Village	15/170 x 120 = 10 people
5	Bajulmati Village	10/170 x 120 = 8 people
6	Staff of Baluran National Park	71/170 x 120 = 50 people
The Total of Sampel		170

Table 1.
Sample size of the Forestry Community Training Center and The Staff of Baluran National Park (BNP).

Variable	Indicators	Outer Loading	Remark
KLM	X_{11}	0.652	Valid
	X_{12}	0.681	Valid
	X_{13}	0.715	Valid
	X_{14}	0.733	Valid
	X_{15}	0.803	Valid
ISN	X_{21}	0.775	Valid
	X_{22}	0.986	Valid
	X_{23}	0.982	Valid
	X_{24}	0.984	Valid
MDS	X_{31}	0.898	Valid
	X_{32}	0.855	Valid
	X_{33}	0.812	Valid
	X_{34}	0.796	Valid
	X_{35}	0.910	Valid
PAR	Z_{11}	0.859	Valid
	Z_{12}	0.915	Valid
	Z_{13}	0.891	Valid
	Z_{14}	0.902	Valid

Table 2.
Convergent validity test.

3.3 Discriminant validity

A discriminant validity test (**Table 3**) was performed after those for convergent validity. It is to identify the validity of instrument items in a model [27]. The discriminant construct validity test will meet the criteria of the discriminant validity if the square roots of AVE are higher than the variable correlation score. KLM (X1) has a square root of AVE 0.793 is more than its correlation 0.669, 0.281, 0.669. ISN (X2) is 0.938, its correlation scores are 0.669, 0.299, and 0.681. MDS (X3) is 0.755, and its correlation scores are 0.281, 0.299, and 0.304. PAR (Z1) is 0.892, and its correlation scores are 0.669, 0.681, and 0.304. The criteria of discriminant validity are therefore met [23, 25]

3.4 Reliability test

Reliability test (**Table 4**) consist of indicator reliability and consistency internal reliability both composite reliability and Cronbach alpha. The reliability test shows that all of the outer loadings are >0.6, and p-value is <0.001 less than 0.05, which means all instruments are reliable [23, 25].

3.5 Consistency internal reliability

Consistency internal reliability was tested both for composite reliability, and Cronbach alpha. The consistency of internal reliability values in this study also more than 0.60 (>0.60), and are thus reliable (**Table 5**).

3.6 Composite reliability

The composite reliability coefficients values in this research are more than 0.70 (see **Table 5**). All variables meet reliability requirements [23]. The value for KLM (forest institution connectedness) is 0.846, ISN (incentive participation program) is 0.967, MDS (social capital) is 0.868, and PAR (public participation) is 0.940.

3.7 Cronbach alpha

Internal consistency test (**Table 6**) can be proved by the exact Cronbach alpha values. The Cronbach alpha are as follows: KLM (X_1) is 0.770, ISN (X_2) is 0.951, MDS (X_3) is 0.809, and PAR (Z_1) is 0.914. The criteria for internal consistency are therefore met [25].

3.8 Indicator reliability

The indicator reliability test (**Table 7**) was done in order to ensure the quality of variable with formative indicators. This test result can be gained from the significant of weights or indicator weights. All of the formative indicators have met the

Correlations among l.vs. with sq. rts. of AVEs					
	Sq. rts of AVE	X1	X2	X3	Z1
KLM (Forest Institution Connectedness)	0.793		0.669	0.281	0.669
ISN (incentive)	0.938	0.669		0.299	0.681
MDS (Social Capital)	0.755	0.281	0.299		0.304
PAR (Public Participation)	0.892	0.669	0.681	0.304	

Table 3.
Discriminant validity test.

Variable	Indicators	Outer Loading	P-value	Remark
KLM (X_1)	X_{11}	0.652	<0.001	Valid
	X_{12}	0.681	<0.001	Valid
	X_{13}	0.715	<0.001	Valid
	X_{14}	0.733	<0.001	Valid
	X_{15}	0.803	<0.001	Valid
ISN (X_2)	X_{21}	0.775	<0.001	Valid
	X_{22}	0.986	<0.001	Valid
	X_{23}	0.982	<0.001	Valid
	X_{24}	0.984	<0.001	Valid
MDS (X_3)	X_{31}	0.898	<0.001	Valid
	X_{32}	0.855	<0.001	Valid
	X_{33}	0.812	<0.001	Valid
	X_{34}	0.796	<0.001	Valid
	X_{35}	0.910	<0.001	Valid
PAR (Z_1)	Z_{11}	0.859	<0.001	Valid
	Z_{12}	0.915	<0.001	Valid
	Z_{13}	0.891	<0.001	Valid
	Z_{14}	0.902	<0.001	Valid

Table 4.
Indicators reliability test.

Composite reliability coefficients				
Variable	KLM	ISN	MDS	PAR
Score	0.846	0.967	0.868	0.940
Remark	Valid	Valid	Valid	Valid

Table 5.
Composite reliability coefficients.

Variable	Croanbach's Alpha	Criteria	Remark
KLM (Forest Institution Connectedness)	0.770	≥0.60	Reliable
ISN (incentive)	0.951	≥0.60	Reliable
MDS (Social Capital)	0.809	≥0.60	Reliable
PAR (Public Participation)	0.914	≥0.60	Reliable

Table 6.
Internal consistency test (Cronbach's alpha of each variable).

requirement of indicator reliability with a p-value less than 0.05 (<0.05) and all instruments are valid [23, 25].

3.9 Collinearity

Variable with formative indicators will meet the requirements of collinearity, if the value of variances inflation factor (VIF) is <3.3. KSJ (Welfare) is a latent variable of welfare with four formative indicators, and has the value of variance inflation factor is 2.527 less than 3.3.

3.10 Partial least square analysis

Goodness of fit (inner model) can be evaluated based on R-squared, adj. R-squared, Cronbach alpha, Avg. Var. Ectrac, full collinearity VIF, and Q-squared value (see **Table 8**). R-squared with high value means the model is good and R-squared can be used for response variable.

The results of R-squared for the public participation (PAR) is 0.979 which means that the contribution of the variables incentive participation program (ISN), social capital (MDS), and welfare (KSJ) to the effect on public participation (PAR) is 97.9%, and the remaining 2.1% is attributable to another variable outside the research model.

Composite reliability value and Cronbach alpha can be used to evaluate research instruments. Based on the output, the composite reliability coefficients are 0.846 for KLM, 0.967 for ISN, 0.868 for MDS, 0.907 for KSJ and 0.940 for PAR. They are more than 0.60 and the Cronbach alpha coefficients are 0.770, 0.951, 0.809, 0.861, and 9.14. All of them are more 0.70 for all variables. Therefore, all variables in this research have met the reliability criteria.

The average variances extracted (AVE) is used to evaluate the discriminant validity, with the criterion that values must be >0.50. The AVE values are as follows: (1) forest institution connectedness (KLM) variable is 0.543; (2) incentive participation program (ISN) variable is 0.880; (2) social capital (MDS) variable is 0.571; (3) welfare (KSJ) variable is 0.713; and (4) public participation (PAR) variable is 0.795. All the variables met the AVE value criterion >0.50 and meet the discriminant validity.

Full collinearity VIFs is a complete collinearity test consisting of vertical and lateral multicollinearity. Lateral collinearity is a collinearity between a predictor latent variable and criteria variables and can be used to test the common method bias. The criterion for the full collinearity test values <3.3. This research has met the full collinearity requirements for all variables; they are 2.228 for KLM, 2.254 for ISN, 1.126 for MDS, and 2.527 for KSJ.

Q-squared is used as a predictive test of the relation between the predictor latent variables and the criterion variables. The Q-squared result can be negative, but the R-squared result must be positive. The estimation result of this output above shows good predictive value; at 0.708 and 0.852, values are more than zero (**Table 8**).

3.11 Loading factor (outer model)

The outer loading values are used to know indicator's weight of every variable. Indicators with high outer loading values show they are strong variable measures (**Table 9**). Forest institutions' connectedness variable is consist of five indicators (accountability, transparency, belief-based relationship, forest rules, and information access) are categorized as not good condition (3.86 < 4.00). The highest outer loading is forest rules (0.803) and means score (4.00) is reflected as good condition. But the lowest mean score is accountability (3.71) is reflected as not good condition and effects the level of public participation.

Variable	Indicators	P-Value	VIF	Remark
KSJ (Welfare)	Y11: household income	<0.001	2.114	Valid
	Y12: household education	<0.001	8.001	Valid
	Y13: household health	0.003	1.441	Valid
	Y14: household supporting facilities	<0.001	6.112	Valid

Table 7.
Indicator reliability test of indicator weights.

Latent variable coefficients	KLM	ISN	MDS	KSJ	PAR
R-squared				0.703	0.979
Adjusted R-squared				0.695	0.979
Composite reliability coefficients	0.846	0.967	0.868	0.907	0.940
Cronbach's alpha coefficients	0.770	0.951	0.809	0.861	0.914
Average variances extracted	0.543	0.880	0.571	0.713	0.795
Full collinearity VIFs	2.228	2.254	1.126	2.527	0.720
Q-squared coefficients				0.708	0.852

Table 8.
Output latent variable coefficients.

Incentive participation program variable consist of four indicators (incentive participation programs of training, agriculture tools, cash payment, and agriculture land use) are categorized as not good conditions (3.81 < 4.00). The highest outer loading is the incentive participation program of agriculture tools (0.986), mean score (3.97), but it is still reflected not good condition (<4.00). The lowest mean score is the incentive participation program of training (3.67) is reflected as not good condition and effects the level of public participation, especially in developing the quality of human resources.

The social capital variable is consist of five indicators (reciprocity, social norms, network interaction, level of trust in the community group, and buffer villages group donations) are categorized as not good conditions (3.93 < 4.00). The highest outer loading is social norms (0.819) and means score is (4.21) is reflected as good condition. But the lowest mean score is buffer village group's donation (3.64) is reflected as not good condition and effects the level of public participation.

Welfare variable is consist of five indicators (household income, household education, household health, and household supporting facilities) are categorized as not good condition (3.81 < 4.00). The highest outer loading is family income (0.877) and mean score (3.86) is reflected as not good condition and effects the level of public participation (<4.00). The lowest mean score is family supporting facilities (3.73) is reflected as not good condition and effects the level of public participation.

Public participation variable is consist of four indicators (participation in planning program, participation in implementation, participation in benefit-sharing, and participation in monitoring and evaluation) are categorized as not good condition with average score is 3.85 or less than 4.00. The highest outer loading is participation in implementation (0.915) and mean score (3.85). The lowest mean score is participation in planning program (3.68) is reflected as not good condition.

3.12 Path coefficients and P values

Path coefficients and p values (**Table 10**) and direct hypothesis (**Table 11**) that: **($H1_a$)** KLM (forest institutions' connectedness) does not have a positive significant effect (0.087) on public participation, with p-value 0.166; **($H1_b$)** KSJ (Welfare) mediates the effect of KLM (forest institution connectedness) on public participation (0.552), with p-value <0.001; **(H2a)** INS (incentive participation program) has shown a positive significant effect (0.196) on public participation (p-value 0.013); **($H2_b$)** welfare (KSJ) mediates the effect of ISN (incentive participation program) (0.273) on public participation with p-value <0.001; **($H3_a$)** MDS (social capital) has a positive significant effect (0.141) on public participation, with p-value 0.056; **($H3_b$)** welfare (KSJ) mediates the effect of MDS (social capital) on public participation (0.177), with p-value 0.023; **(H4)** welfare (KSJ) has a positive significant effect (0.782) on public participation with p-value <0.001.

Variables	Indicators	*Outer Loading*	Mean of Indicator	Mean of Variable
Forest Institution Connectedness (X1)	Accountability (X_{11})	0.614	3.71	**3.86**
	Transparancy (X_{12})	0.687	3.81	
	Belief Based Relationship (X_{13})	0.642	3.78	
	Forest Rules (X_{14})	**0.803**	**4.00**	
	Information Access (X_{15})	0.733	3.99	
Incentive (X2)	Incentive of Training ($X_{2.1}$)	0.775	3.67	**3.81**
	Incentive of Agriculture Tools ($X_{2.2}$)	**0.986**	**3.97**	
	Incentive of Cash Payement ($X_{2.3}$)	0.982	3.78	
	Incentive of Agriculture Land Using (X_{24})	0.984	3.83	
Social Capital (X3)	Reciprocity ($X_{3.1}$)	0.753	3.98	**3.93**
	Social Norms ($X_{3\text{-}2}$)	**0.819**	**4.21**	
	Network Interaction ($X_{3\text{-}3}$)	0.620	3.88	
	Level of Trust to the Community Group ($X_{3\text{-}4}$)	0.780	3.93	
	Buffer Vaillages Group's Donation ($X_{3\text{-}5}$)	0.789	3.64	
Welfare (Y1)	Family Income ($Y_{1\text{-}1}$)	**0.866**	**3.86**	**3.81**
	Family Edution ($Y_{1\text{-}2}$)	0.664	3.82	
	Family Healty ($Y_{1\text{-}3}$)	0.812	3.85	
	Family Supporting Facilities ($Y_{1\text{-}4}$)	0.636	3.73	
Public Participation (Z1)	Participation in planning program ($Z_{1.1}$)	0.859	3.68	**3.85**
	Participation in implementation ($Z_{1.2}$)	**0.915**	**4.03**	
	Participation in benefit-sharing ($Z_{1.3}$)	0.891	3.81	
	Participation in evaluation and monitoring ($Z_{1.4}$	0.902	3.88	

Table 9.
Outer loading value of variable.

Path coefficients					
	KLM	ISN	MDS	KSJ	PAR
KLM					
ISN					
MDS					
KSJ	0.552	0.273	0.177		
PAR	0.087	0.196	0.141	0.782	
P values					
	KLM	ISN	MDS	KSJ	PAR
KLM					
ISN					
MDS					
KSJ	<0.001	<0.001	0.023		
PAR	0.166	0.013	0.056	<0.001	

Table 10.
Path coefficients and P values.

3.13 Directional hypothesis

Forest institutions' connectedness does not have a positive significant effect (0.087) on public participation, with p-value 0.166. Because p-value 0.166 is more than 0.05 (0.166 > 0.05), $\mathbf{H_{1a}}$ is not accepted. This test result does not provide empirical support for the findings of Baynes et al. [7], Muro and Namusonge [8], and Lise [9].

The incentive participation program has a positive significant effect (0.196) on public participation, with p-value 0.013. Because p-value 0.013 is less than 0.5

(0.013 < 0.05), $\mathbf{H_{2a}}$ is accepted. The results support the theory of incentive participation programs of Robbin [28], Adhikari et al. [12], Djamhuri [13], and Kaseya and Kihonge [14].

Social capital gives significant positive effect on public participation (0.1401), with p-value 0.056. Because p-value 0.056 is less than 0.05 (0.056 < 0.5), $\mathbf{H_{3a}}$ is accepted. The test results support the theory of social capita [29], as well as supporting the empirical research of Sara [18] and Sharpe [19].

3.14 Indirect effect hypothesis

Path coefficient indirect effect (**Table 12**) shows that welfare mediates the effect of forest institution on public participation (0.552), with p-value <0.001. Because p-value <0.001 is less than 0.05 (<0.001 < 0.5), hypothesis $\mathbf{H_{1b}}$ is accepted. The test results provide empirical support for the work of Hans-Jurgen [10] and Akib et al. [11].

Welfare mediates incentive participation program on public participation (0.273), with p-value < 0.001, less than 0.05 (<0.001 < 0.5). Hypothesis $\mathbf{H_{2b}}$ is thus accepted. The test results provide empirical support for the work of Rahut et al. [15], William and Ayuk [16], Das and Sarker [17].

Social capital by the mediation of welfare has a positive significant effect on public participation (0.177), with p-value 0.023, less than 0.05 (0.023 < 0.5). Hypothesis $\mathbf{H_{3b}}$ is therefore accepted. The results are related to the social capital theory [20]. In addition, Fukuyama [30] added that the social capital and the level of welfare are closely related in a community or nation [29]. This result provides empirical support for the research of Grootaet [20], Narayan and Pritchett [21].

Welfare contributes significant positive effect on public participation by 0.782 on public participation, with p-value <0.001. Because p-value <0.001 is less than 0.05 (<0.001 < 0.5), hypothesis $\mathbf{H_4}$ is accepted. This test provides empirical support for the research of Rahut et al. [15] and Akamani and Hall [22].

3.15 Mediation effect analysis

To test mediation effect, this research uses Baron and Kenny's causal-step approach. Baron and Kenny [26] using causal step approach which has four mediation effects, they are: (a) first step, directional hypothesis if the results are significant/positive; (b) second step, the indirect hypothesis was tested whether it is significant/positive; (c) third step, test mediation effects using VAF (Variance Accounted For) with the criteria: VAF value >80% means full mediation, 20% ≤ VAF ≤80% means partial mediation; and VAF < 20% means no mediation. The mediation effect is significant/positive if p-value indirect effect is less than 0.05 [25].

Figure 1 shows that all of the direct effects are significant/positive because the p-values are less than 0.05. Then the indirect effects (mediation variables) are included, as shown in **Figure 2**.

Hypothesis	Direct Effect Among Latent Variables			Path Coefficient Value	P-value	Remark
	Independent Variable	To	Dependent Variable			
H1a	KLM	→	PAR	0.087	0.166	Not Significant
H2a	ISN	→	PAR	0.196	0.013	Significant
H3a	MDS	→	PAR	0.141	0.056	Significant

Table 11.
Direct hypothesis.

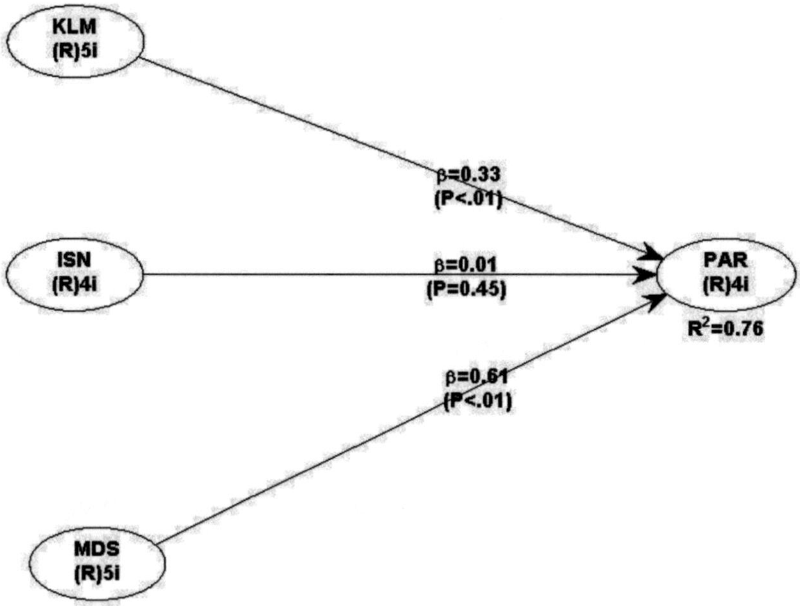

Figure 1.
Direct effect without including mediation.

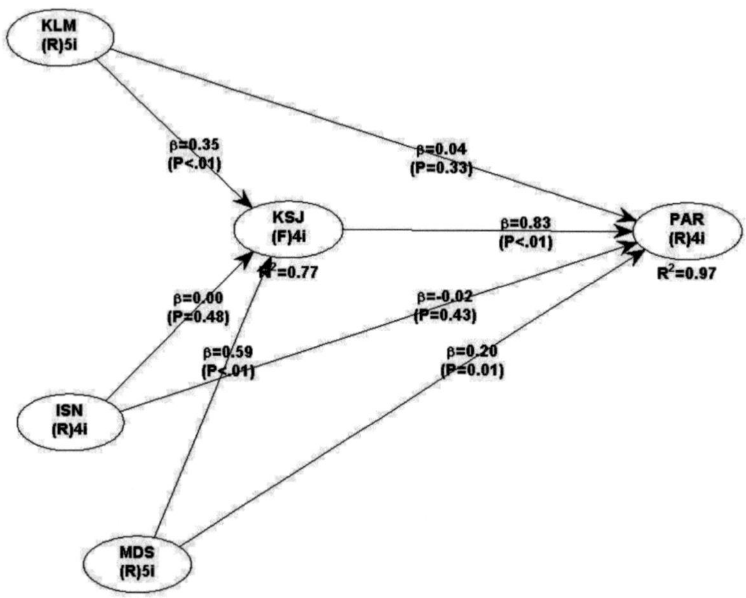

Figure 2.
Indirect effect including mediation.

Figure 2 shows all of the indirect effects are significant/positive because p-values are less than 0.05. The third step is to test mediation effect by using the VAF formula. The formula of VAF = (p12 × p23)/(p12 × p23 + p13). The results of the mediation test using the VAF method are as follows:

1. Forest institutions' connectedness effect on public participation in the mediation of welfare is significant and positive with p-value <0.001 (<0.05)

VAF = (0.552 × 0.782)/(0.552 × 0.782 + 0.087).
VAF = 0.431/0.518.
VAF = 0.832.
VAF = 83.2%.

This means that welfare mediates the effect of forest institutions' connectedness on public participation as a full mediation.

2. Incentive participation program's effect on public participation in the mediation of welfare is significant and positive, with p-value <0.001 (<0.05)

VAF = (0.273 × 0.782)/(0.273 × 0.782 + 0.196).
VAF = 0.213/0.409.
VAF = 0.520.
VAF = 52.0%.

This means that welfare mediates incentive participation program's effect on public participation as a partial mediation.

3. Social capital's effect on public participation in the mediation of welfare is significant and positive, with p-value 0.023 (<0.05)

VAF = (0.177 × 0.782)/(0.177 × 0.782 + 0.141).
VAF = 0.138/0.279.
VAF = 0.494.
VAF = 49.4%.

This means welfare mediates social capital's effect on public participation as a partial mediation.

Based on the descriptive analysis, both direct and indirect hypothesis results (**Tables 11** and **12**), for all variables can be summarized in **Table 13**.

3.16 Analysis of public participation and welfare as mediator of forest management

The analysis of public participation in this research is based on the characteristics of five buffer villages. They are Wonorejo, Sumber Waru, Sumber Anyar, Watu Kebo and Bajul Mati.

The buffer villages have potential to be developed into bigger villages. Management of regions is required in order to avoid disturbing the forest conservation in Baluran National Park.

The contribution of nontimber forest product (NTFP) to family income is around 19.79% up and 61.44% of their total annual income (**Table 14**).

Hypothesis	Indirect Effect Among Latent Variables			Path Coefficient Value	P-value	Remark
	Independent Variable	To	Dependent Variable			
H1b	KLM	PAR	KSJ	0.552	<0.001	Significant
H2b	ISN	PAR	KSJ	0.273	<0.001	Significant
H3b	MDS	PAR	KSJ	0.177	0.023	Significant
4	KSJ	→	PAR	0.782	<0.001	Significant

Table 12.
Path coefficient indirect effect.

Hypothesis	Direct Effect Among Latent Variables Independent Variable	To	Dependent Variable	Path Coefficient Value	P-value	Remark
H1a	KLM	→	PAR	0.087	0.166	Not Significant
H2a	ISN	→	PAR	0.196	0.013	Significant
H3a	MDS	→	PAR	0.141	0.056	Significant
Hypothesis	**Indirect Effect Among Latent Variables Independent Variable**	**To**	**Dependent Variable**	**Path Coefficient Value**	**P-value**	**Remark**
H1b	KLM	PAR	KSJ	0.552	<0.001	Significant
H2b	ISN	PAR	KSJ	0.273	<0.001	Significant
H3b	MDS	PAR	KSJ	0.177	0.023	Significant
4	KSJ	→	PAR	0.782	<0.001	Significant

Table 13.
Hypothesis test results, summary of direct and indirect effect.

This research also empirically supports research from Baluran [5] and Syafi'i [31]. The total income of user's forest product in buffer villages of Baluran National Park is about Rp. 100,900,000 a year.

3.17 Public participation in the forestry planning program in Baluran National park

The people in the buffer villages have not been widely involved in the forestry management planning (shown in **Table 15**). Their participation in decision making in the meeting of the forestry planning program by forestry community training center is 28.2%, and by Forest Institution of Baluran National Park is 20.6%.

3.18 Public participation in the implementation program

The members of Forestry Community Training Center participated by giving inputs of the forestry planning program in BNP (23.5%) and giving efforts and actions of the forestry planning program (17.6%) as shown in **Table 16**.

No	Name fo Villages	Agriculture's Income	Non Timber Forest's Product (NTFP)	Cattle's Business Income	Total	Contribution of NTFP (%)
1	Wonorejo	18,500,000	7,500,000	11,900,000	37,900,000	19.79
2	Bajul Mati	4,400,000	3,500,000	2,000,000	9,900,000	35.35
3	Watu Kebo	7,900,000	21,350,000	5,500,000	34,750,000	61.44
4	Sumber Anyar	3,100,000	3,450,000	2,300,000	8,850,000	38.98
5	Sumber Waru	3,300,000	3,800,000	2,400,000	9,500,000	40.00
	Total	37,200,000	39,600,000	24,100,000	100,900,000	

Table 14.
Contribution of nontimber forest product (NTFP) on family income.

No	Public Participation of Forestry Planning Program	Level of Participation High f	%	Middle f	%	Low f	%	Total f	%
1	Members of Forestry Community Training Center receive invitation of meeting to make decision held by BNP management	48	28.2	67	39.4	55	32.35	170	100.0
2	Members of Forestry Community Training Center have opportunities to give suggestions to make forestry planning program	35	20.6	60	35.3	75	44.12	170	100.0

Table 15.
Public participation in the forestry planning program in Baluran National Park.

3.19 Public participation in benefit sharing

The members of the Forestry Community Training Center have participated in the benefit-sharing from forestry management to increase their family's income with value 21.8%, and have participated in forest conservation management with value 17.1% (**Table 17**).

3.20 Public participation in evaluation and monitoring

Table 18 shows that 17.6% of members of Forestry Community Training Center has participated in the evaluation of Baluran Forest in Baluran National Park, and 22.4% have participated in the monitoring of forest conservation.

3.21 Classification of public participation in forestry management in Baluran National Park

According to Cohen [32], the level of public participation is high when people involved in four stages of the management process. They are (1) program planning participation; (2) actuating participation; (3) benefit-sharing participation; and (4) evaluation and monitoring participation. Scores > 21 indicate high level, 17–21 medium level, and <17 low-level participation [32].

No	Public Participation in Implementing Program	Level of Participation							
		High		Middle		Low		Total	
		f	%	f	%	f	%	f	%
1	Members of Forestry Community Training Center have Participated to give inputs of the forestry planning program in BNP	40	23.5	85	50.0	45	26.5	170	100.0
2	Members of Forestry Community Training Center have Participated to give efforts, and actions) of the forestry planning program in BNP	30	17.6	75	44.1	65	38.2	170	100.0

Table 16.
Public participation in implementing program.

No	Public Participation of Benefit Sharing	Level of Participation							
		High		Middle		Low		Total	
		f	%	f	%	f	%	f	%
1	Member of Forestry Community Training Center have participated in gaining of benefit from forestry management to increase family	37	21.8	87	51.2	46	27.1	170	100.0
2	Member of Forestry Community Training Center have participated in the benefit of forest conservation management	29	17.1	93	54.7	48	28.2	170	100.0

Table 17.
Public participation of benefit sharing.

No	Public Participation of Evaluation and Monitoring	Level of Participation							
		High		Middle		Low		Total	
		f	%	f	%	f	%	f	%
1	Member of Forestry Community Training Center have participated in evaluation of Baluran forest in the Baluran National Park	30	17.6	86	50.6	54	31.8	170	100.0
2	Member of Forestry Community Training Center have participated in monitoring of forest conservation in the Baluran National Park	38	22.4	82	48.2	50	29.4	170	100.0

Table 18.
Public participation in evaluation and monitoring.

No	Level of Public Participation Category	Name of Buffer Villages					Total of Respondent	Percentage (%)
		Wonorejo	Sumber Waru	Sumber Anyar	Bajul Mati	Watu Kebo		
1	Low Participation (Score <17)	46	16	11	2	5	80	47.1
2	Middle Participation (Score 17-21)	34	12	9	1	1	57	33.5
3	High Participation (Score >21)	15	7	5	3	3	33	19.4
	Total	**95**	**35**	**25**	**6**	**9**	170	

Table 19.
Classification of public participation in forestry management in Baluran National Park.

Table 19 shows that public participation from the five buffer villages (Wonorejo, Sumber Waru, Sumber Anyar, Bajul Mati, and Watu Kebo) is low level at 47.1%, the middle level at 33.5%, and high level 19.4%.

4. Conclusions

Based on the research findings, several conclusions can be stated as follows. Public welfare is the most important factor in forest management. Especially in the Baluran National Park, that the public welfare is the main factor affecting public participation. That is why, the public participation will increase if: (1) there is a good relationship between the forest institution connectedness and local people and make them welfare and better in their life than before; (2) the forest institution provide incentive participation program that can increase the local people's welfare and better in their life; (3) there is a good social capital that can increase the local people's welfare because they are more having skills, experiences and productivity; (4) the public welfare always increase and make their life better; (5) public welfare is the most important factor to increase public participation in forest management; (6) the contribution of nontimber forest product (NTFP) to family income getting increased; (7) the forest institutions give opportunities to member of forestry community training center in the forestry planning program, implementing program, benefit sharing, and evaluation and monitoring. Classification of public participation in forestry management in the Baluran National Park shows that the public participation needs to be increased (19.4%).

5. Policy recommendations

Based on the findings of this study, the following recommendations are proposed:

1. The forest institution of Baluran National Park should improve its relationship between the institution connectedness and local people. Because of this, the institution should know what the local people needs.

2. The forest institution should increase the incentive participation program (incentive participation program of training), because it effects the level of public participation;

3. The social capital should be improved to create the local people's welfare by giving incentive participation program of training in order to improve the skills, experiences and productivity of local people;

4. The forest institution of Baluran National Park should improve the public welfare by focusing it on each planning program of forest management;

5. The contribution of nontimber forest product (NTFP) to family income must be increased;

6. The forest institutions give opportunities to member of forestry community training center in the forestry planning program, implementing program, benefit sharing, and evaluation and monitoring. Classification of public participation in forestry management in the Baluran National Park shows that the public participation needs to be increased (19.4%).

7. Future research should develop socio-demography variable as a predictor variable of public participation in optimizing the level of public participation as many researchers using it in their study.

Acknowledgements

We first thank Indonesia Endowment Fund for Education and Training Ministry of Finance (LPDP), The Republic of Indonesia funding my study in doctoral management and business, University of Brawijaya. We thankful also to the professionals involved in forest management in the Baluran National Park, and the local key informants, who shared their views and experiences for this research.

Author details

Adil Siswanto[1,2]* and Djumilah Hadwidjojo[1]

1 Doctoral in Management and Business, Faculty of Economic and Business, University of Brawijaya, Malang, Indonesia

2 Hotel Accomodation Teacher State Vocational High School (SMKN 2 Bondowoso), East Java, Indonesia

*Address all correspondence to: adil_siswanto@yahoo.com

References

[1] Dirjen. Pengelolaan Pemberdayaan Masyarakat Di Daerah Penyangga (Vol. Dipa BA-29 Tahun 2008): Satker Direktorat Pemanfaatan Jasa Lingkungan dan Wisata Alam Direktorat Jenderal PHKA; 2008

[2] Tahajuddin U. Pengelolaan Sumber Daya Hutan: Suatu Tantangan. Pengelolaan Hutan Lestari: Partisipasi, Kolaborasi, dan Konflik. Jakarta: Yayasan Pustaka Obor Indonesia; 2015

[3] Muntasib H. Hutan dan Lingkungan, Kerjasama Pusat Penyuluhan Kehutanan dan Perkebunan: Fakultas Kehutanan Institut Pertanian Bogor/IPB; 1999

[4] Baluran TN. Rencana Pengelolaan Jangka Panjang Taman Nasional Baluran Tahun 2014–2023. Situbondo: Direktorat Jenderal Kehutanan dan Konservasi Alam, Kementerian Kehutanan; 2014

[5] Baluran TN. Laporan Inventarisasi Hasil Hutan Non Kayu yang Dimanfaatkan Masyarakat Serta Penyebaranyya pada Taman Nasional Baluran. Situbondo: Direktorat Jenderal Perlindungan Hutan dan Pelestarian Alam, Balai TN. Baluran, Banyuwangi; 1997

[6] Sabarno MY. Savana di Taman Nasional Baluran. Jurnal Biodiversitas. 2001;**3**(1):207-221

[7] Baynes J, Herbohn J, Smith C, Fisher R, Bray D. Key factors which influence the success of community forestry in developing countries. The Journal of Global Environmental Change. 2015;**35**. DOI: 10.1016/j.gloenvcha.2015.09.011

[8] Muro JE, Namusonge GS. Governance factors affecting community participation in public development projects in Meru District in Arusha in Tanzania. Journal of Scientific & Technology Research. 2015;**4**(06)

[9] Lise W. Factors influencing people's participation in forest management in India. Journal of Ecological Economics. 2000;**34**:379-392

[10] Hans-Jurgen W. Good governance, welfare, and transformation. The European Journal of Comparative Economics. 2004;**1**(1):127-143

[11] Akib M, Habbe AH, Rura Y, Hakim A. Relationship good governance with welfare society mediated by local government financial disclosure and audit report opinion (Study on Local Government South Southeast Sulawesi). International Journal of Science and Research (IJSR). 2016;**5**(2)

[12] Adhikari S, Tanira K, Siva G. Incentive for community participation in the governance and management of common property resources: The case of community forest management in Nepal. Journal of Forest Policy and Economic. 2014;**44**(9). DOI: 10.1016/j.forpol.2014.04.003

[13] Djamhuri TL. The effect of incentive structure to community participation in a social forestry program on state forest land in Blora District, Indonesia. The Journal of Forest Policy and Economics. 2012;**25**(18). DOI: 10.1016/j.forpol.2012.02.004

[14] Kaseya CN, Kihonge E. Factors affecting the effectiveness of public participation in county governance in Kenya: A case of Nairobi county. International Journal of Scientific and Research Publications. 2016;**6**(10)

[15] Rahut DB, Akhter A, Bhagirath B. Household public participation and effects of community forest management on income and poverty levels: Empirical evidence from Bhutan. Journal of Forest Policy and Economics. 2015;**61**

[16] William MF, Ayuk ET. Measuring the role of forest income in mitigating poverty and inequality: Evidence from South Eastern Nigeria. Journal of Forests, Trees and Livelihoods. 2013

[17] Das N, Sarker D. Joint forest management program: A beneficial relationship between state and community. International Journal of Joint Forest Management Program. 2010

[18] Zare S, Namiranian M, Shabanali FH, Ghasemi J. The role of social capital on citizen's participation in the management of forest parks (Case Study: Tehran City). Iranian Journal of Forest. 2011;**2**(4):273-285

[19] Sharpe B. First the forest': Conservation, 'community' and 'participation' in South-West cameroon. Journal of the International African Institute. 2012;**68**(1):25-45

[20] Grootaet. Capital, household walfare and poverty in Indonesia. Paper presented at the local level institutions; Washington, DC; 1999

[21] Narayan D, Pritchett L d. Cents and sociability: Household income and social capital in rural Tanzania. Journal of Economic Development and Cultural Change. 1997;**47**(4):871-897

[22] Akamani K, Hall TE. Determinants of the process and outcomes of household participation in collaborative forest management in Ghana: A quantitatif test of a community resilience model. Journal of Environmental and Management. 2015; **147**(1–11). DOI: 10.1016/j. jenvman.2014.09.007

[23] Kock N. WarpPLS 5.0. User Manual. Texas, USA: Laredo, TX: Script Warp System; 2015

[24] Creswell JW, Plano Clark VL, Gutmann ML, Hanson WE. Advanced Mixed Methods Research Designs. Thousand Oaks, CA: Sage Publications; 2003

[25] Hair JF Jr, Hult GTM, Ringle CM, Sarstedt M. A Primer on Partial Least Squares Structural Equation Modeling (PLS-SEM). London, United Kingdom: SAGE Publications; 2014

[26] Baron RM, Kenny DA. The moderator-mediator variable distinction in social psychological research: Conceptual, strategic, and statistical considerations. Journal of Personality and Social Psychology. 1986;**51**(6): 1173-1182

[27] Zikmund WG, Babin BJ, Carr JC, Griffin M. Business Research Method. Vol. 9. South-Western-USA: M. Roche Ed. Cengage Learning; 2013

[28] Robbin S. Perilaku Organisasi. Jilid 1 Edisi Bahasa Indonesia. Jakarta: Salemba Empat; 2007

[29] Woolcock M. Social Capital and Economic Development; Toward a Theoretical Synthesis and Policy Framework. Theory and Society. In: Elinor O, Ahn TK, editors. Foundation of Social Capital. Massachusetts: Edward Elgar Publishing Limited; 1998

[30] Fukuyama. Social capital and civil society. Paper presented at the Paper presented at the IMF Conference on Second Generation Reforms, Washington, DC; 1999

[31] Syafi'i A, Manikasari GP, Wistantama HA, Janiawati IAA, Satria D, Ryan A. Laporan Hasil Praktik Pengelolaan Kawasan Konservasi Resort Labuhan Merak Taman Nasional Baluran. Yogyakarta; 2013

[32] Cohen U. Rural Development Participation: Concept and Measures for Project Design Implementation and Evaluation Rural Development Participation. New York: Cornel University Press; 1997

Chapter 3

Prescribed Burning to Restore Eastern White Pine Forests of La Mauricie National Park of Canada

Christian Hébert, Éric Domaine and Louis Bélanger

Abstract

Eastern white pine forests of La Mauricie National Park of Canada have been severely affected by logging and forest fire suppression since the 1850s, and by the exotic white pine blister rust since the beginning of the twentieth century. These alterations have changed the ecological trajectory of eastern white pine ecosystems, which now appear hardly sustainable. Eastern white pine saplings are nearly absent, and balsam fir saplings are strong competitors for space and light. Since 1991, Parks Canada uses prescribed burning for restoring eastern white pine ecosystems. We studied seven pine stands in which prescribed burning was applied and compared them with nine unburned stands. Over 63% of balsam fir saplings were killed by prescribed burning, thus eliminating a significant part of the competition to eastern white pine seedlings. These were four times more abundant in burned than in unburned sites (21,333 vs. 5178 seedlings/ha). In the short term, the eastern white pine regeneration objectives established by Parks Canada have been achieved. Pine seedlings growth is slow, and they should be monitored regularly to ensure long-term success of this restoration programme. If necessary, it might be helpful to increase light penetration by girdling mature balsam firs or spruces.

Keywords: *Pinus strobus*, eastern white pine, prescribed burning, ecosystem restoration, protected area, regeneration, competition, *Abies balsamea*, balsam fir

1. Introduction

Over the last decades, changes in forest composition from primeval stages and the rarefaction of certain tree species have raised several concerns for biodiversity conservation [1, 2]. For example, the widespread mortality of ash trees caused by the alien invasive Emerald ash borer (*Agrilus planipennis* Fairmaire), recently introduced in North America, could threaten 43 native arthropod species feeding or breeding only in ash trees [3]. Indeed, exotic insect pests and pathogens may lead to tree species shifts and be a driving force behind important changes in ecosystem processes [4]. This already occurred in western North America where the exotic white pine blister rust, *Cronartium ribicola* J.C. Fisch., introduced at the beginning of the twentieth century, affects seven of the eight white pines (subgenus *Strobus*) [5]. For instance, the whitebark pine (*Pinus albicaulis* Engelmann) has been extirpated locally because of the combined actions of the white pine blister rust, an indigenous insect, and fire suppression policies [6]. These authors consider the whitebark pine as a foundation species because it provides locally stable conditions needed by several co-occurring species. Its loss thus alters several ecosystem processes such as forest productivity and hydrology [6]. According to Tomback and Achuff [5], without active management, many pine-associated communities may disappear and their loss would result in severe impacts to biodiversity and other ecosystem services. They recommend using timely proactive restoration programmes to avoid or at least mitigate losses in pine ecosystems.

In eastern North America, the eastern white pine (*Pinus strobus* L.) was also much more prevalent in pre-settlement forests than it is today [7–9]. Eastern white pine is the tallest tree in eastern North America and ecologically typifies the northern forests of eastern United States [10]. This noble tree species has been important for economic, social, and cultural reasons [11]. As western white pines, it has been also severely impacted by the exotic white pine blister rust, *Cronartium ribicola* J.C. Fisch [12]. Moreover, fire suppression policies have altered the natural dynamics of eastern white pine stands by allowing shade-tolerant species, such as balsam fir (*Abies balsamea* L.), to outcompete pine seedlings [8, 10, 13–15]. Finally, selective logging of mature eastern white pines during the eighteenth and nineteenth centuries has reduced seed tree density and, thus, its regeneration potential [16, 17].

Historically, eastern white pine regeneration was favoured by surface fires, which improve seedbed quality, increase light availability, and reduce competition from saplings of other shade-tolerant tree species [7, 14, 18]. Mature eastern white pines survive most surface fires due to their thick bark [19], branch-free lower trunks, and deep roots [20]. Their needles have a low content of resin and thus are not highly flammable [21]. Eastern white pine reaches the northern limit of its range in the southern part of eastern Canada, where most ignitions are rapidly suppressed for safety reasons. Other than fire, the natural regeneration dynamics of eastern white pine in old stands is still poorly understood, mainly at the northern limit of its range [11]. Recent studies have recognized the importance of gap dynamics, which is closely related to understory light for seedlings [22, 23]. Uprety et al. [11] concluded that management strategies should be different near the northern range limits because site conditions and disturbances have different effects than in the centre of a species' range. Regenerating eastern white pine thus remains an important challenge and researchers still test methods to reduce the effect of competing vegetation [24]. However, this mainly involves using herbicides [24] or thinning [25], approaches not compatible with the mandate of national parks.

The *Canada National Parks Act* requires maintaining or restoring the ecological integrity of the parks through the protection of natural resources and ecological processes. Ecological integrity is defined as 'a condition that is determined to be characteristic of its natural region and likely to persist, including abiotic components and the composition and abundance of native species and biological communities, rates of change and supporting processes' [26]. Because several parks have been established in areas previously disturbed by logging, Parks Canada often needs to develop management approaches to restore these ecosystems to make them sustainable for future. The objective of restoring the ecological integrity of eastern white pine forest ecosystems to pre-settlement conditions, or at least within their historic range of variability [27, 28], might be achieved by using prescribed burning as a management approach [29, 30]. In the context of a national park, prescribed burning represents a tool for reintroducing a natural ecological process. Prescribed burning has been shown to promote regeneration of several fire-favoured pine species, such as *P. ponderosa*, *P. pungens*, and *P. rigida* [31–34], but its efficacy remains to be demonstrated in eastern white pine forests.

In La Mauricie National Park of Canada, eastern white pine proportion was estimated at 5–12% in pre-settlement forests but now represents only 0.5% of the current forest composition. Meanwhile, balsam fir has increased from 13.1 to 31.8% [35–37].

Between 1991 and 2005, a total of 10 Eastern white pine stands have been treated with prescribed burning in La Mauricie National Park of Canada. The objectives of prescribed burning are to generate ecological conditions for increasing eastern white pine seedling density (short-term objective) to bring saplings density up to 100/ha (mid-term objective) in order to increase the cover of eastern white pine-dominated stands to 3–4% (long-term objective) in the future forested area of the park [15]. As the number of published studies addressing the ecological effectiveness of management practices in protected areas is limited [38], this restoration programme represents a unique opportunity to evaluate the effects of prescribed burning, as a sustainable management practice for a national park. The objective of this study was to determine if prescribed burning reduces competition and favours eastern white pine regeneration. We hypothesized that prescribed burning would kill most balsam fir competing saplings, thus reducing competition for light and promoting eastern white pine regeneration.

2. Materials and methods

2.1 Study area, stand selection, and burn treatment

The study was carried out in La Mauricie National Park of Canada (**Figure 1**), which is located in Quebec, Canada. The park was established in 1977 and covers 536 km^2. It belongs to the sugar maple-yellow birch bioclimatic domain and is a typical Laurentian Mountains landscape moulded with hills and lakes. Annual precipitations vary between 900 and 1400 mm and annual mean temperatures vary between 2.5 and 5.0°C [39]. Seven stands treated with prescribed burning between 1995 and 2005, and nine unburned stands, were selected over an area of 40 km^2. The altitude of the 16 selected stands ranged between 217 and 341 m and their slope varied between 1 and 47% (**Table 1**).

Prescribed burning was used in stands where eastern white pine density was >15 trees/ha, the slope <50%, and balsam fir saplings dominated the understory. In these sites, eastern white pine seedling and sapling densities were considered too

Figure 1.
Old white pine stand along the Wapizagonke lake in La Mauricie National Park of Canada.

Site	Fire intensity	Humus (cm)	Altitude (m)	Slope (%)
Burn 2005	3	2.7	242	4
Burn 2004	3	4.3	245	2
Burn 2003	3	3.7	341	41
Burn 2001	3	5.3	286	26
Burn 1999	4	6.7	217	5
Burn 1997	2	7.3	250	4
Burn 1995	2	6.7	262	1
Control 1	-	6.0	259	5
Control 2	-	9.0	314	38
Control 3	-	9.3	303	47
Control 4	-	6.7	220	1
Control 5	-	6.7	277	46
Control 6	-	5.3	249	9
Control 7	-	4.3	264	16
Control 8	-	5.7	259	18
Control 9	-	9.3	239	9

Table 1.
Description of the 16 sites studied at La Mauricie National Park of Canada.

Figure 2.
Parks Canada crew using a driptorch to run a prescribed burning experiment in a white pine stand at La Mauricie National Park of Canada.

low to ensure eastern white pine renewal, thus preventing the park from reaching its objective of maintaining or restoring ecological integrity [40]. Burning prescriptions were defined using the Canadian Forest Fire Danger Rating System [41] and the software FBP97 for forecasting fire behaviour [42]. Prescribed burnings were carried out during spring because burning conditions are more suitable before bud flushing of broadleaved trees and shrubs [43]. When conditions were appropriate, fire was ignited using burners (driptorch; **Figure 2**) or a helicopter equipped with a Premo MK3 aerial ignition device. Low-intensity surface fires were isolated and controlled with natural and artificial firebreaks. Flame height and length were recorded during each prescribed burning event by the park's staff and were used to estimate fire intensity based on the Canadian forest fire behaviour prediction system [41] (**Table 1**). For low-intensity surface fires, these classes range from 1 (frontal fire intensity < 10 kW/m; flame length < 0.2 m; flame height < 0.1 m) to 5 (frontal fire intensity > 4000 kW/m; flame length >3.5 m; flame height >2.5 m). In our study, fire intensity in burned sites mostly belongs to class

3 (frontal fire intensity: 500–2000 kW/m; flame length: 1.4–2.6 m; flame height: 1.0–1.9 m). However, fire intensity reached class 4 in the stand burned in 1999 (frontal fire intensity: 2000–4000 kW/m; flame length: 2.6–3.5 m; flame height: 1.9–2.5 m) and killed many mature trees including some eastern white pines.

2.2 Forest inventory

Three 400-m^2 circular plots located 50 m apart along a transect and at a minimum distance of 50 m from stand or treatment edges were set up in each stand to describe the forest environment. In each plot, we recorded the slope (%), altitude (m), surface deposit, drainage, and thickness of the soil organic layer (litter and humus) (**Table 1**).

Species, diameter at breast height (hereafter DBH), and decay class of each standing tree or snag ≥9.1 cm at DBH were recorded. Decay classes were determined according to Hunter classification [44], which recognizes nine classes for trees (1: alive and 2: declining) and snags (3: dead tree with bark intact up to 9: stump). Because most pines were large and tall, their density was rather low and, to get more accurate estimates of their basal area, we enlarged the sampled plots up to 1200 m^2 (radius = 19.55 m). In each 400-m^2 plot, four smaller plots of 25 m^2 (radius = 2.82 m) and four micro plots of 4 m^2 (radius = 1.13 m) were established at 8.46 m from the plot centre, in each cardinal direction. Saplings and seedlings were recorded in the 25 and 4-m^2 plots, respectively. Saplings were defined as young trees in which DBH ranged between 1 and 9 cm, whereas seedlings were very young trees with DBH smaller than 1 cm [45]. Sapling DBH was measured and seedling height was recorded into 5-cm classes. Eastern white pine relative dominance was estimated on the basis of its relative basal area (hereafter BA, in m^2/ha) in 1200-m^2 plots, in relation to BA of other tree species estimated in the 400-m^2 plots.

2.3 Statistical analysis

As stands had not been sampled before treatment, the short-term effects of prescribed burning were assessed using the percentage of recent tree or sapling mortality in 1- to 7-year-old burns (older burns could not represent short-term effects of prescribed burning) and compared to unburned stands. Tree BA and sapling density (stems/ha) were calculated for eastern white pine, balsam fir, spruces, and broadleaved species. Then, the percentages of recent mortality (Hunter classes 3 and 4) were calculated for both burned and unburned stands. Student's t-tests were used to compare recent mortality of trees and saplings in both stand types. We also used Student's t-tests to compare seedling density in burned and unburned stands. Sites burned in 2004 and 2005 were excluded from the seedling analysis because no seed crop had occurred after the treatment, thus precluding the establishment of regeneration in these stands. Logarithmic transformations (log x + 1) were used to normalize the distributions and stabilize variances when necessary. When transformations did not achieve equality of variances, we used results obtained with Satterthwaite's approximate t-test, a method that belongs to the Behrens-Welch family [46]. Analyses were performed using SAS software v. 9.1. [47].

3. Results

Forest composition of unburned stands was dominated by conifers, with slightly more than 75% of the tree basal area belonging to eastern white pine and other conifers, mostly spruces (**Table 2**). Balsam fir represented less than 10% of tree basal area, and broadleaved trees slightly more than 15%. Prescribed burnings significantly

Stage	Species	Initial BA[1] or density[2] of live trees		% of recent mortality		df	t	p
		Burned	Unburned	Burned	Unburned			
Tree	*P. strobus*	13.7 ± 2.8	11.4 ± 2.0	1.4 ± 0.8	2.4 ± 1.5	14	0.52	0.611
	A. balsamea	3.1 ± 0.7	3.1 ± 1.1	38.2 ± 12.7	3.7 ± 2.1	14	3.20	**0.006**
	Other conifers	14.6 ± 2.7	18.1 ± 2.9	18.4 ± 6.8	10.7 ± 2.8	14	1.07	0.304
	Broadleaves	6.1 ± 1.3	5.8 ± 1.2	39.0 ± 8.8	14.3 ± 3.3	14	2.81	**0.014**
	Total	37.8 ± 3.4	38.1 ± 2.2	18.8 ± 4.5	11.3 ± 1.8	14	1.65	0.120
Saplings	*P.strobus*	9 ± 6	26 ± 11	0.0 ± 0.0	0.0 ± 0.0	-	-	-
	A.balsamea	2590 ± 372	2396 ± 427	67.4 ± 7.3	9.2 ± 2.6	14	4.28	**0.001**
	Otherconifers	214 ± 144	437 ± 132	43.4 ± 18.0	4.1 ± 2.0	14	1.37	0.192
	Broadleaves	238 ± 57	189 ± 44	37.0 ± 11.7	6.1 ± 3.2	14	2.42	**0.030**
	Total	3052 ± 333	3048 ± 391	63.6 ± 6.9	9.2 ± 2.3	14	4.46	**0.001**

[1] basal area for trees (Burned sites = 7 / Unburned sites = 9)
[2] density for saplings (Burned sites = 5 / Unburned sites = 9)

Table 2.
Comparison of average (± S.E.) initial tree basal area and sapling density (before burning) and of the % of recent mortality of different species in burned and unburned stands. Student t-tests were used to compare mortality averages between treatments (significant ones are in bold).

increased the mortality of balsam fir and broadleaved trees, with respectively 38.2 and 39.0% compared with 3.7 and 14.3% in unburned stands (**Table 2**). Mature eastern white pine trees and other conifers were not significantly affected by the burning treatment, except in the 1999 burn which was the most intense (**Table 1**). Eastern white pines were well distributed among DBH classes, with maximum densities recorded between 20 and 60 cm of DBH (**Figure 3A** and **B**). However, small balsam fir trees (≤20 cm) as well as saplings outnumbered eastern white pines (**Figure 3C** and **D**). Balsam fir saplings represented 80% of total sapling density while eastern white pine represented only 0.9% (**Table 2**).

Saplings of balsam fir and broadleaved species were significantly affected by the burning treatment (**Table 2; Figure 4**). Mortality averaged 67.4 and 37.0% respectively for balsam fir and broadleaved saplings in burned stands compared with 9.2 and 6.1% in unburned stands (**Table 2**). The most severely burned stand (1999) had killed 93% of the balsam firs, which was 25% higher than in any other burned stand. Overall, mortality of saplings was significantly higher in burned stands (63.6%) than in unburned ones (9.2%) (**Table 2**). However, after burning, the density of balsam fir saplings was still high, mostly because patches of the forest remained unburned in some stands (**Figure 3C**).

Eastern white pine seedling density was lower than for balsam fir in each height class observed in unburned stands (**Figure 3F**) and they represented only 26.7% of all seedlings (**Table 3**). Moreover, they never reached more than 75 cm in height (**Figure 3F**). Prescribed burnings increased the density of eastern white pine seedlings significantly when compared with unburned stands (**Table 3**), their proportion increasing from 26.7 to 83.7% of all seedlings in burned stands. By contrast, the proportion of balsam fir seedlings decreased from 39.6% in unburned stands to 20.6% in burned ones (**Table 3**). However, eastern white pine seedlings were largely dominant in the first five height classes (1–25 cm), but they rarely exceeded 50 cm. Balsam fir seedlings were more evenly distributed up to 130 cm and dominated eastern white pine seedlings in all height classes higher than 25 cm (**Figure 3E**).

No distinctive pattern in eastern white pine seedling growth was obvious along the burning chronosequence. Stands burned in 2004 and 2005 only harboured 417 seedlings of eastern white pine per hectare because these sites (1–2 years after burning) had not yet benefited from a good seed production year [48, 49].

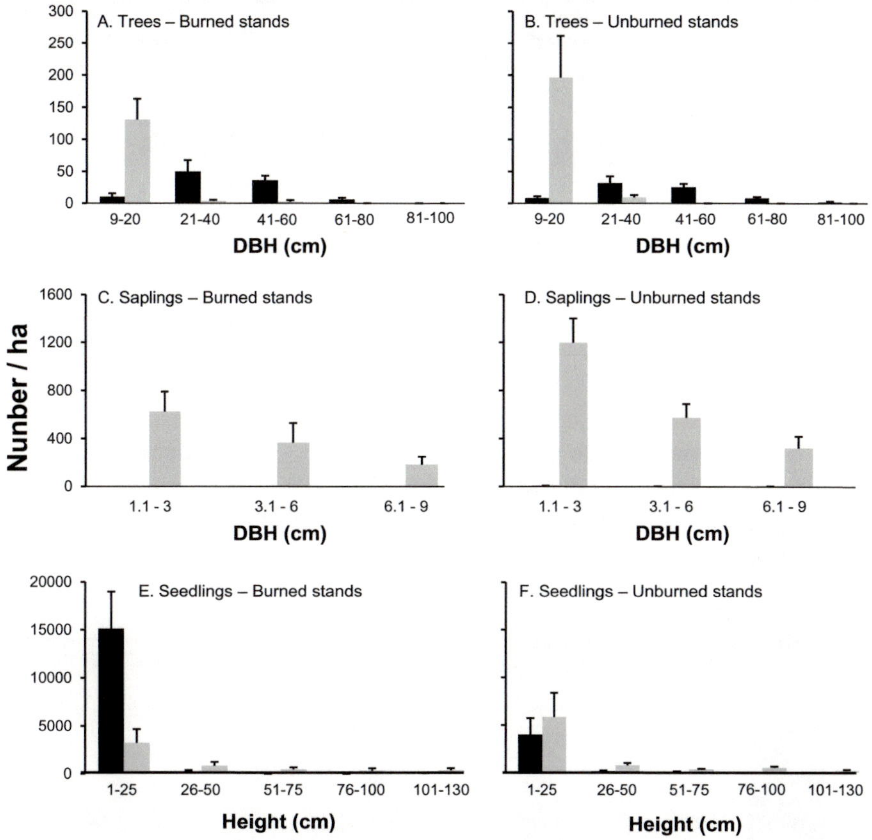

Figure 3.
Comparisons of eastern white pine (black bars) and balsam fir (grey bars) tree (A and B), sapling (C and D), and seedling (E and F) densities (mean ± S.E.) between burned (n = 7) and unburned white pine stands (n = 9) of La Mauricie National Park of Canada.

Figure 4.
Photo showing abundant competing balsam fir seedlings and their reduction 1 year after a prescribed burning in an eastern white pine stand at Lac Guilinette of La Mauricie National Park of Canada.

Stage	Species	Density / ha (mean ± SE) Burned	Unburned	p	Proportion (%) Burned	Unburned
Seedlings	*A. balsamea*	5250.0 ± 1869.7	7592.9 ± 2451.8	0.543	20.6	39.6
	P. strobus	21333.0 ± 1368.7	5135.8 ± 1716.6	<0.0001	83.7	26.7
Saplings	*A. balsamea*	746.4 ± 250.6	2017.2 ± 296.1	0.016	52.0	80.1
	P. strobus	6.6 ± 6.6	24.2 ± 9.7	0.255	0.3	0.9

Table 3.
Comparison of living sapling and seedling densities (after burning) of Eastern white pine and balsam fir in burned (n = 5; 2004–2005 excluded) and unburned sites (n = 9).

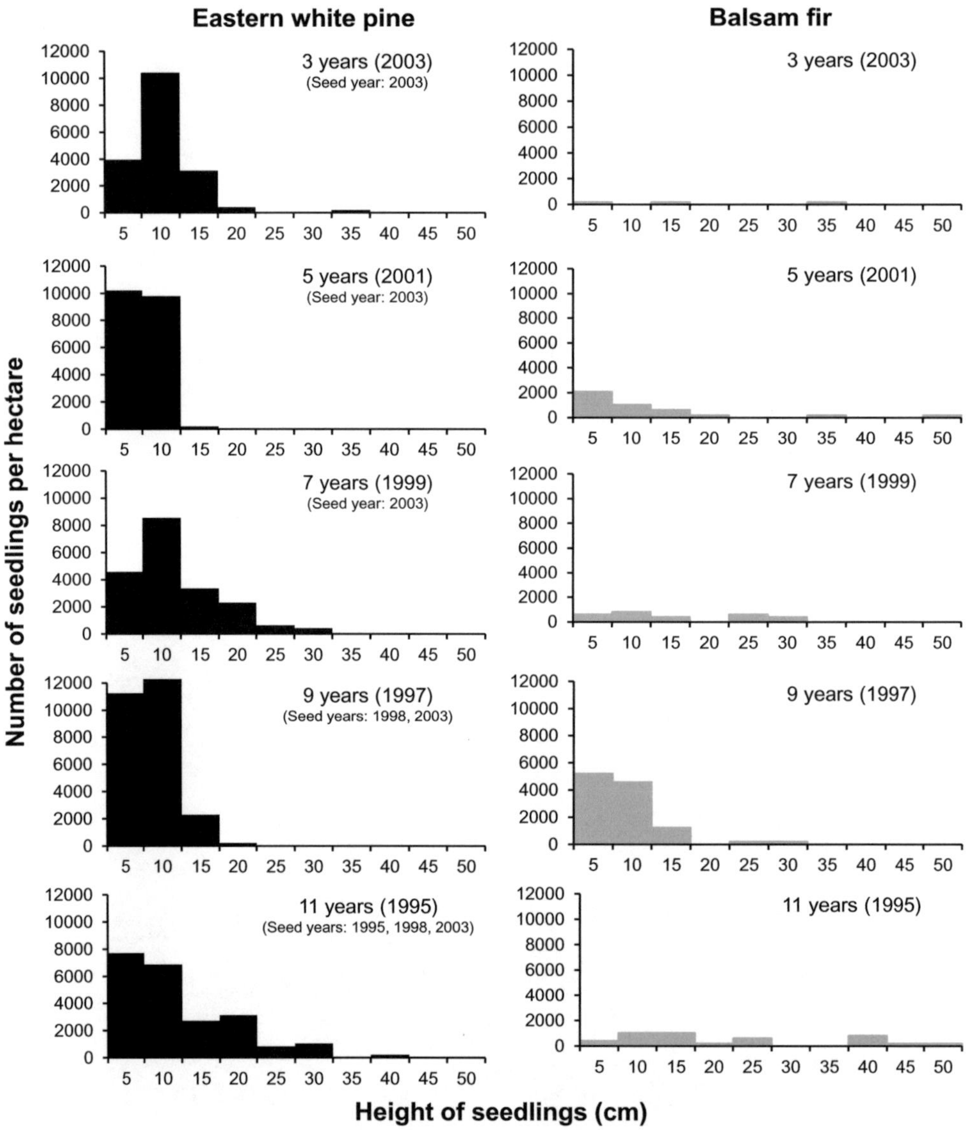

Figure 5.
Distribution of eastern white pine and balsam fir seedlings among 5-cm height classes in burned stands of various ages in La Mauricie National Park of Canada. No eastern white pine seedling exceeded 45 cm of height.

The stand burned in 2003 (3 years) was exposed to a good seed crop the summer after the treatment and regeneration was the most plentiful in the 5- to 10-cm height class. The stand burned in 2001 (5 years) was exposed to a good seed crop 2 years after the treatment and seedlings were slightly more abundant than in the stand burned in 2003, with almost all seedlings being found in the 5- to 10-cm height class (**Figure 5**). In the stand burned in 1999 (7 years), eastern white pine seedlings were asymmetrically distributed to the right of the 5- to 10-cm height class, with seedlings reaching 25–30 cm (**Figure 5**). However, the stand burned in 1997 (9 years), which benefited from two good seed crops (1998, 2003), did not harbour more eastern white pine seedlings than the other stands, with its seedlings being mainly found in the first two height classes (1–5 cm and 5–10 cm) (**Figure 5**). In the oldest burned stand (1995; 11 years) that benefited from three good seed crops, seedlings were mainly found in the first two height classes. However, eastern white pine seedlings reached their maximum height (40–45 cm) in this stand. Balsam fir seedlings <50 cm high were not abundant in any of these five burned stands compared with eastern white pine seedlings, the only one showing more than 10,000 balsam fir seedlings/ha being the one that burned in 1997 (**Figure 5**).

4. Discussion

Prescribed burnings carried out in La Mauricie National Park of Canada killed 38 and 67% of competing balsam trees and saplings respectively and increased eastern white pine seedling density up to an average of 21,133 seedlings/ha, compared with 5135 seedlings/ha in unburned stands. This is higher than the 12,000 seedlings/ha reported 5 years after small scale (0.2 ha) prescribed burning trial done in an eastern white pine stand at the Petawawa forest research station located in Ontario [41]. Stands burned in 2004 and 2005 had very low density of eastern white pine seedlings when compared with other burned stands because no seed crop occurred after the treatment was applied and the time of our study. However, in these stands, the bracken fern species (*P. aquilinum*) was abundant with an average cover of 26%. This fern is highly competitive in recolonizing burned stands when it is present prior to treatment. It is a fire-adapted species that possesses deep fire-resistant rhizomes [50]. Bracken ferns are strong competitors for light and their presence is known to increase the level of competition for the establishment of eastern white pine seedlings [13]. In the future, after a first good seed crop, it would be important to monitor and measure the regeneration of eastern white pine and the effect of the abundance of this fern on pine dynamics. The production of a good seed crop is important after fire or a burning treatment in order to establish strong pine regeneration. Moreover, the succession of various events may also be favourable to establish pine regeneration as described by Lynham and Curran [51]. They reported 50,000 red and white pine seedlings per hectare 5 years after a low-intensity natural fire followed by a good seed crop 2 years later and a blowdown 4 years after the fire. This generated optimal conditions for regenerating pines. Such condition may explain why one of our unburned stand, which was located on an island, had a high density of eastern white pine saplings (100/ha) even though its seedling density was rather low (625/ha). This stand had been disturbed by a small blowdown due to its location on an island. Gaps produced in the forest cover were large enough to favour the growth of eastern white pine seedlings up to the sapling stage. This is important for eastern white pine forest renewal as Stiell [52] demonstrated that pine's ability to compete is greatly improved when the sapling stage is reached. However, even if all the saplings recorded in our unburned stands would reach the canopy, their numbers would still remain below the density objective of 100/ha [40].

The stands burned in 2001 and 2003 benefited from a good seed crop in 2003 and they had numerous seedlings, approximating 20,000 seedlings/ha. Most pine seedlings were grouped in the 5- to 10-cm height class and showed a normal growth rate for eastern white pine [13]. The stand burned in 1999 was submitted to the highest fire intensity and showed the highest recent mortality of trees, including some pines. It also produced taller eastern white pine seedlings than the site burned 2 years before. No competing vegetation reinvaded this stand and the good 2003 seed crop allowed regenerating eastern white pine, even if it occurred only 4 years after the treatment. Furthermore, the taller seedlings observed indicate better growth conditions in this stand. This could be linked not only with better soil conditions but also with a higher penetration of light due to higher tree and sapling mortality. Finally, the tallest eastern white pine seedlings were observed in the oldest burned site, but these only reached 50 cm of height, which shows a slow growth rate after 11 years [13].

5. Conclusions

Active management is an important approach for restoring the ecological integrity of ecosystems in Canadian national parks. The current policy states that when park ecosystems have been seriously altered by human activities and natural processes cannot achieve restoration objectives alone, intervention may be prescribed. In La Mauricie National Park of Canada, the ecological integrity of eastern white pine forest ecosystems has been altered by logging, fire suppression, and the introduction of the exotic white pine blister rust. On a short-time scale (10–15 years), the prescribed burning programme implemented in the park has been successful in increasing eastern white pine seedling density significantly. However, in the near future, it would be important to continue monitoring each burned stand to make sure that local environmental conditions remain favourable for the growth of eastern white pine seedlings. Initial growth of eastern white pine usually averages 10–15 cm after 5 years [13], which is slow compared with faster growing competitors such as firs and hardwoods. In order to evaluate the ecological integrity of a national park, Timko and Innes [53] recently recommended such monitoring for assisting managers in evaluating the effectiveness of their management actions.

According to Ontario Ministry of Natural Resources [54], eastern white pine seedlings that receive more than 45% of full light have a higher probability of reaching the sapling stage. Otherwise, seedlings will probably survive but might not be able to grow rapidly enough to outcompete firs and broadleaved species. Waldrop and Brose [31] have shown that low-intensity prescribed burnings in Ponderosa pine stands do not open sufficiently the overstory strata to ensure survival and fast growth of seedlings established after treatment. In such cases, it might be necessary to use further treatment to reach the objective of restoring pine forest ecosystems. It might be the case in burned stands of La Mauricie National Park of Canada as densities of eastern white pine seedlings are high, but their growth appears rather slow. Eastern white pine seedlings may benefit from opening the canopy to increase light penetration. In Ponderosa pine stands, it has been shown that thinning was more effective than burning to open the overstory and kill a higher proportion of mature trees [33, 34]. These authors concluded that the combination of thinning and burning was the most effective option for optimizing light penetration and ensuring good seedling growth. In eastern white pine stands of La Mauricie National Park of Canada, we may hypothesize that this objective could be achieved by girdling mature balsam fir or spruce trees or by increasing fire intensity in future prescribed burnings. Girdling appears as a better option than thinning for a national Park as it leaves large diameter snags on site which increases the treatment value with respect to the concept of ecological integrity. Prescribed burning effectiveness for killing mature trees can be enhanced by increasing fire intensity but, this also increases risks of escaping the fire, which represents a fragile equilibrium between the ecological value of the treatment and safety rules [34]. Nevertheless, continuous monitoring of seedling growth in burned stands would help managers to confirm that such management practices are useful in La Mauricie National Park of Canada to restore eastern white pine ecosystems. These evaluations would also promote adaptive management and ensure that decision-making is based on sound science [55].

Acknowledgements

We thank Yves Dubuc from Natural Resources Canada as well as the summer students Jennifer Naples, Félix-Antoine Boucher, Jonathan Boucher, Éloïse Laurier, and Vincent Mandon for their help with field work. We also acknowledge Victor Kafka, Raymond Quenneville, and Michel Thériault from Parks Canada for their logistical support.

Funding

Financial support to run this project was provided by Parks Canada and Natural Resources Canada.

Author details

Christian Hébert[1*], Éric Domaine[2,3] and Louis Bélanger[2]

1 Natural Resources Canada, Québec, Canada

2 Université Laval, Québec, Canada

3 Ministère des Forêts, de la Faune et des Parcs, Québec, Canada

*Address all correspondence to: christian.hebert@canada.ca

References

[1] Spies TA, Franklin JF. The diversity and maintenance of old-growth forests. In: Szaro RC, Johnston DW, editors. Biodiversity in Managed Landscapes: Theory and Practice. New York: Oxford University Press; 1996. pp. 296-314. Chapter 20

[2] Tiscar PA, Lucas-Borja ME. Structure of old-growth and managed stands and growth of old trees in a mediterranean *Pinus nigra* forest in southern Spain. Forestry. 2016;**89**:201-207. DOI: 10.1093/forestry/cpw002

[3] Gandhi KJK, Herms DA. North American arthropods at risk due to widespread *Fraxinus* mortality caused by the alien emerald ash borer. Biological Invasions. 2010;**12**:1839-1846. DOI: 10.1007/s10530-009-9594-1

[4] Lovett GM, Canham CD, Arthur MA, Weathers KC, Fitzhugh RD. Forest ecosystem responses to exotic pests and pathogens in Eastern North America. BioScience. 2006;**56**:395-405. DOI: 10.1641/0006-3568(2006)056[0395,FERTEP]2.0.CO;2

[5] Tomback DF, Achuff P. Blister rust and western forest biodiversity: Ecology, values and outlook for white pines. Forest Pathology. 2010;**40**:186-225. DOI: 10.1111/j.1439-0329.2010.00655.x

[6] Ellison AM, Bank MS, Clinton BD, Colburn EA, Elliott K, Ford CR, et al. Loss of foundation species: Consequences for the structure and dynamics of forested ecosystems. Frontiers in Ecology and the Environment. 2005;**3**:479-486. DOI: 10.1890/1540-9295(2005)003[0479,LOFSCF]2.0.CO;2

[7] McRae DJ, Lynham TJ, Frech RJ. Understory prescribed burning in red pine and white pine. The Forestry Chronicle. 1994;**70**:395-401. DOI: 10.5558/tfc70395-4

[8] Carleton TJ, Maycock PF, Arnup R, Gordon AM. In situ regeneration of *Pinus strobus* and *P. resinosa* in the Great Lakes forest communities of Canada. Journal of Vegetation Science. 1996;**7**:431-444. DOI: 10.2307/3236287

[9] Doyon F, Bouffard D. Enjeux écologiques de la forêt feuillue tempérée québécoise, Québec. Pour le ministère des Ressources naturelles et de la Faune, Direction de l'environnement et de la protection des forêts. 2009. p. 63. Available from: http://www.mrnf.gouv.qc.ca/forets/amenagement/amenagement-ecosystemique.jsp

[10] Abrams MD. Eastern white pine versatility in the pre-settlement forest. BioScience. 2001;**51**:967-979. DOI: 10.1641/0006-3568(2001)051[0967:EWPVIT]2.0.CO;2

[11] Uprety Y, Asselin H, Bergeron Y, Mazerolle MJ. White pine (*Pinus strobus* L.) regeneration dynamics at the species' northern limit of continuous distribution. New Forests. 2016;**45**:131-147. DOI: 10.1007/s11056-013-9396-2

[12] Kinloch BB Jr. White pine blister rust in North America: Past and prognosis. Phytopathology. 2003;**93**:1044-1047. DOI: 10.1094/PHYTO.2003.93.8.1044

[13] Horton KW, Bedell GHD B. White and Red Pine: Ecology, Silviculture and Management. Ottawa, Canada: Canadian Department of Northern Affairs and Natural Resources, Forestry Branch; 1960. Bulletin 124

[14] Methven IR, Murray WG. Using fire to eliminate understory balsam fir in pine management. The Forestry Chronicle. 1974;**50**:77-79. DOI: 10.5558/tfc50077-2

[15] Quenneville R, Thériault M. Cadre pour la restauration écologique du pin

blanc au Parc National de la Mauricie. Shawinigan, Canada: Parcs Canada, Service de la conservation des ressources naturelles; 1998

[16] Palik BJ, Pregitzer KS. White pine seed-tree legacies in an aspen landscape: Influences on post-disturbance white pine population structure. Forest Ecology and Management. 1994;**67**:191-201

[17] Smidt MF, Puettmann KJ. Overstory and understory competition affect underplanted eastern white pine. Forest Ecology and Management. 1998;**105**:137-150. DOI: 10.1016/S0378-1127(97)00278-8

[18] Maissurow DK. Fire as a necessary factor in the perpetuation of white pine. Journal of Forestry. 1935;**33**:373-378. DOI: 10.1093/jof/33.4.373

[19] Hengst GE, Dawson JO. Bark properties and fire resistance of selected tree species from the central hardwood region of North America. Canadian Journal of Forest Research. 1994;**24**:688-696. DOI: 10.1139/x94-092

[20] Farrar JL. Les Arbres du Canada. Saint-Laurent, Canada: Service canadien des forêts, Ressources Naturelles Canada et Fides; 1995

[21] Landers JL. Disturbance influences on pine traits in the southeastern United States. In: Proceedings of the 17th Tall Timbers fire ecology conference; 18-21 May 1989; Tallahassee, FL; 1991. pp. 61-95

[22] Leithead M, Silva LCR, Anand M. Recruitment patterns and northward tree migration through gap dynamics in an old-growth white pine forest in northern Ontario. Plant Ecology. 2012;**213**:1699-1714. DOI: 10.1007/s11258-012-0116-3

[23] Parker WC. The relationship of stand structure with canopy transmittance: Simple models and practical methods for managing understory light conditions in eastern white pine (*Pinus strobus* L.) dominated forests. The Forestry Chronicle. 2014;**90**:489-497. DOI: 10.5558/tfc2014-099

[24] Santala K, Aubin I, Hoepting M, Bachand M, Pitt D. Managing conservation values and tree performance: Lessons learned from 10 year experiments in regenerating eastern white pine (*Pinus strobus* L.). Forest Ecology and Management. 2019;**432**:748-760. DOI: 10.1016/j.foreco.2018.09.038

[25] Burgess D, Wetzel S, Pinto F. Regenerating eastern white pine: A cooperative research approach. The Forestry Chronicle. 1999;**75**:423-425. DOI: 10.5558/tfc75423-3

[26] Department of Justice Canada. Canada National Parks Act (2000, c. 32) Ottawa, Canada. 2000. Available from: http://laws.justice.gc.ca/en/N-14.01/index.html

[27] Landres PB, Morgan P, Swanson FJ. Overview of the use of natural variability concepts in managing ecological systems. Ecological Applications. 1999;**9**:1179-1188. DOI: 10.1890/1051-0761(1999)009[1179:OOTUON]2.0.CO;2

[28] Brown RT, Agee JK, Franklin JF. Forest restoration and fire: Principles in the context of place. Conservation Biology. 2004;**18**:903-912. DOI: 10.1111/j.1523-1739.2004.521_1.x

[29] Barden LS, Woods FW. Effects of fire on pine and pine-hardwood forests in the southern Appalachians. Forest Science. 1976;**22**:399-403. DOI: 10.1093/forestscience/22.4.399

[30] Harrod RJ, McRae BH, Hartl WE. Historical stand reconstruction in ponderosa pine forests to guide silvicultural prescriptions.

Forest Ecology and Management. 1999;**114**:433-446. DOI: 10.1016/S0378-1127(98)00373-9

[31] Waldrop TA, Brose PH. A comparison of fire intensity levels for stand replacement of table mountain pine (*Pinus pungens* Lamb.). Forest Ecology and Management. 1999;**113**:155-166. DOI: 10.1016/S0378-1127(98)00422-8

[32] Welch NT, Waldrop TA, Buckner ER. Response of southern Appalachian table mountain pine (*Pinus pungens*) and pitch pine (*P. rigida*) stands to prescribed burning. Forest Ecology and Management. 2000;**136**:185-197. DOI: 10.1016/S0378-1127(99)00291-1

[33] Youngblood A, Metlen KL, Coe K. Changes in stand structure and composition after restoration treatments in low elevation dry forests of northeastern Oregon. Forest Ecology and Management. 2006;**234**:143-163. DOI: 10.1016/j.foreco.2006.06.033

[34] Harrod RJ, Peterson DW, Povak NA, Dodson EK. Thinning and prescribed fire effects on overstory tree and snag structure in dry coniferous forests of the interior Pacific Northwest. Forest Ecology and Management. 2009;**258**:712-721. DOI: 10.1016/j.foreco.2009.05.011

[35] Richard P. La vulgarisation des travaux paléobiogéographiques effectués dans le parc national de la Mauricie (Rapport final). Service de la recherche en analyse pollinique. Chicoutimi, Canada: Université du Québec à Chicoutimi; 1975

[36] Pelletier H. Plan de conservation des écosystèmes terrestres. Shawinigan, Canada: Parcs Canada, Parc National du Canada de la Mauricie, Service de la conservation des ressources naturelles; 1998

[37] Barrette M. Caractérisation du paysage primitif de la région écologique des hautes collines du Bas-Saint-Maurice pour une gestion des écosystèmes du parc national du Canada de la Mauricie (Mémoire de maîtrise). Québec: Université Laval; 2004. p. 129

[38] Gaston KJ, Charman K, Jackson SF, Armsworth PR, Bonn A, Briers RA, et al. The ecological effectiveness of protected areas: The United Kingdom. Biological Conservation. 2006;**132**:76-87. DOI: 10.1016/j.biocon.2006.03.013

[39] Gosselin J. Guide de reconnaissance des types écologiques de la région écologique 3c—Hautes collines du Bas-Saint-Maurice, 2e édition. Ministère des Ressources naturelles, Direction des inventaires forestiers, Division de la classification écologique et productivité des stations; 2014

[40] Quenneville R, Thériault M. Plan de gestion du feu: Parc National du Canada de la Mauricie. Shawinigan, Canada: Parcs Canada, Parc National du Canada de la Mauricie, Service de la conservation et des écosystèmes; 2002

[41] Forestry Canada Fire Danger Group. Development and Structure of the Canadian Forest Fire Behaviour Prediction System. Forestry Canada Information Report ST-X-3;1992

[42] Remsoft Inc. Software FBP97. Remsoft, Fredericton, Canada; 1997

[43] Vlasiu PD, Nolet P, Doyon F. Le pin blanc: revue de littérature. Ripon, Canada: Institut québécois d'aménagement de la forêt feuillue; 2001

[44] Hunter ML. Wildlife, Forests and Forestry: Principles of Managing Forests for Biological Diversity. Lebanon, USA: Prentice Hall Inc.; 1990

[45] Canada. Silvicultural terms in Canada. Science and Sustainable Development Directorate, Forestry Canada; 1992. p. 63

[46] Armitage P, Berry G, Matthews JNS. Statistical Methods in Medical Research. 4th ed. Oxford, U.K.: Blackwell Science Ltd; 2001. p. 832

[47] SAS Institute Inc. SAS for Windows: Version 9.1. Cary, USA: SAS Institute Inc.; 2000

[48] Frelich LE. The relationship of natural disturbances to white pine stand development. In: Stine RA, Baughman MJ, editors. Proceedings of White Pine Symposium: History, Ecology, Policy and Management. St. Paul, USA: University of Minnesota; 1992. pp. 27-37

[49] Nolet P, Doyon F, Vlasiu PD. L'aménagement par coupe par trouées des strates de feuillus d'essences tolérantes avec pin blanc. Ripon, Canada: Institut québécois d'aménagement de la forêt feuillue; 1999

[50] Gliessman SR. The establishment of bracken following fire in tropical habitats. American Fern Journal. 1978;**68**:41-44. DOI: 10.2307/1546778

[51] Lynham TJ, Curran TR. Vegetation recovery after wildfire in old-growth red and white pine. Natural Resources Canada, Frontline Express Bulletin No. 31. 2003

[52] Stiell WM. Silviculture of eastern white pine. Proceedings of the Entomological Society of Ontario. 1985;**116**(Suppl):95-107

[53] Timko JA, Innes JL. Evaluating ecological integrity in national parks: Case studies from Canada and South Africa. Biological Conservation. 2009;**142**:676-688. DOI: 10.1016/j.biocon.2008.11.022

[54] Ontario Ministry of Natural Resources. Ecology and Management of Eastern White Pine in the Lake Abitibi (3E) and Lake Temagami (4E) Ecoregions of Ontario. Sault Ste. Marie, Canada: Ontario Ministry of Natural Resources; 2009

[55] Pullin AS, Knight TM, Stone DA, Charman K. Do conservation managers use scientific evidence to support their decision-making? Biological Conservation. 2004;**119**:245-252. DOI: 10.1016/j.biocon.2003.11.007

Chapter 4

Institutionalizing Co-Management for a Sustainable Future of Protected Areas: The Case of Xuan Thuy National Park, Vietnam

Nguyen Kim Dung

Abstract

To conserve nature and biodiversity, Vietnam has established 164 protected areas, comprising of 30 national parks, 58 nature conservation areas, 10 species and habitat reserves, 46 land/seascape protected areas, and 20 scientific and experimental forest areas. Like many other developing countries, Vietnam has been facing many institutional challenges to govern the system. Insufficiencies of human and financial resources, conflicts over customary and statutory laws, overlaps of land use rights, and deficiencies in legitimate rights and responsibilities are those complicate the situation. To overcome the obstacles, the state needs supports from multilevel government, community, and international, private, and civil societies. Co-management has been suggested and implemented as a form of governance that can help mobilize the engagement of diversified stakeholders as well as harmonize conflicts over the areas. However, transformation from a centralized governance like Vietnam to a co-management requires time and effort; it reveals a promising process for a sustainable future of the Vietnamese protected areas through some initial achievement.

Keywords: protected areas, nature conservation, institution, co-management, Vietnam

1. Introduction

Located within the Indo-Burma Biodiversity Hotspot (IBBH), Vietnam is ranked as the 16th most biodiversity-rich country in the world. It hosts 110 Key Biodiversity Areas [1] and 62 Important Bird Areas [2]. The country also claims two World Natural Heritage sites, eight Ramsar wetlands, eight United Nations Education, Scientific and Cultural Organization (UNESCO) biosphere reserves, and two Association of Southeast Asian Nations (ASEAN) heritage parks. Underlying this list of conservation governance arrangements is the country's high level of species endemism. It is estimated that 10% of Vietnam's plants are endemic to the country [3], while 12 known species of mammals, 7 species of birds, 48 species of reptiles, 33 species of amphibians, and 80 species of freshwater fish are endemic to Vietnam [4].

To conserve nature and biodiversity, Vietnam has established 164 protected areas, comprising of 30 national parks, 58 nature conservation areas, 10 species and habitat reserves, 46 land/seascape protected areas, and 20 scientific and experimental forest areas [5, 6]. The legal basis of the protected system is the restriction of resource exploitation which can adversely affect biodiversity, natural and cultural landscapes, and scientific resources (Decree 117/2010/ND-CP). Under the rigid set of policies and law enforcement, the protected area system has been evaluated as not supportive of local livelihoods [7–10], although about 80% of the protected areas are inhabited [11]. This leads to the exclusion of the people out of the system since its establishment [12] and degrading relations between local people and the protected areas [13, 14].

To manage the system, the government plays the sole role. At the national level, the Vietnam Forestry Administration (VFA), within the Ministry of Agricultural and Rural Development (MARD), is primarily responsible for coordinating the national protected area system, including the direct administration of six inter-provincial national parks [15, 16]. Where other habitats and resources involved, protected area management can also involve other branches of the government. For instance, wetlands are divided between the Ministry of Natural Resources and Environment (MONRE) and MARD [17]. MARD remains the main authority responsible for marine protected areas [18]. But if cultural or landscape protection is involved in a protected area, then the Ministry of Culture, Sports and Tourism is also involved.

At the provincial level, the Provincial People's Committees (PPCs) are responsible for the administration of all other protected areas [12]. Based on the size and importance of forests, PPCs might assign district governments to manage and develop activities such as tourism [19]. But the majority of Vietnam's protected areas is managed by the provincial Departments of Agriculture and Rural Development (DARDs), in collaboration with the provincial Forest Protection Department of Culture, Sports, and Tourism (DOCST) which fall directly under the control of the PPCs [16, 20]. Protected areas that are small in size and not managed by the district-level Forest Protection Department report directly to PPCs. At the operational level, management boards of protected areas are staffed by officials assigned by provincial DARDs and responsible for management and protection [21]. Due to this fragmentation of institutional arrangement, it requires a lot of efforts placed on the coordination for achieving the effectiveness in protected area management [22]. As acknowledged by the Vietnamese government itself, the coordination between the authorities is plagued by overlapping legislation and a lack of clear division between institutional mandates for management by the various authorities involved [4, 5, 16].

Although the relative large number of protected areas has been established, there is a continuous decrease in quality of forests, biodiversity, and wildlife habitats because the government faces shortages in human and financial resources to carry out the management [12]. Beside deficiencies in institutions, it has been dealing with big challenges in nature conservation when there are millions of people still directly or indirectly depending on these protected areas for their livelihoods [23]. According to the Government (2014), about 20 million people in Vietnam have main or partial income from aquatic resources and 20–50% of income of 25 million people from non-timber forest products [24]. This is fueled with the issues of population growth putting more pressures on resource exploitation and socioeconomic development, threatening Vietnam's natural resources. Moreover, overlapping land use rights is another central issue when 49% of protected areas remain dealing with conflicts over the ambiguities that arise over access to both land and forest resources. Relations between managers, rangers, and local communities still struggled over livelihood and conservation and between statutory and customary laws ([25], p. 11). Conflicts between local users and authorities responsible for nature conservation commonly happen when the former have been forcefully excluded from protected areas [26].

A combination of poor surveillance and weak active engagement with local communities depending on the resources causes problems for protected areas. McElwee [27] argued that extending the protected network areas is not feasible in the case of Vietnam because of the limited capacity of institutions. In recognition of the same issues, the Vietnamese government has attempted through a variety of programs to improve community collaboration in the protected area system and improve the protection of forest habitat and biodiversity [28]. Co-management approach has strong potentials to resolve the issues as suggested by international scholars when it is supposed to provide a meaningful participation through joint decision-making [29–31], a means of conflict resolution [32], and a reduction in resource management cost with more locally relevant management plans for poverty reduction through diversifying economic activities [33, 34]. And thus it is exactly going to contribute to the sustainable development of Vietnam by ways of integrating nature conservation and development.

2. Concepts of sustainable development and natural resource co-management

2.1 Sustainable development

Sustainability originated with the 1980 World Conservation Strategy of the International Union for Conservation of Nature (IUCN). It is considered as a strategic approach to the integration of conservation and development consistent with the objectives of ecosystem maintenance, the preservation of genetic diversity, and the sustainable utilization of resources. In general, "sustainable development is development that meets the needs of the present, without compromising the ability of future generations to meet their own needs" (World Commission on Environment and Development, 1987). Today, sustainable development remains a controversial topic with different philosophies ranging from resource conservation to socioeconomic equality through environmental justice [35]. Although it views a holistic approach to the relationship between man and the environment, various actors emphasize differing facets of sustainable development depending on their needs. Moreover, cultural and economic differences also lead to different perceptions of sustainable development [35].

Recently, sustainable development is understood and related to four key aspects, comprising of environment, development, society, and linkages, among poverty, inequality, and environmental degradation. The environment composing of nature with biophysical domains and human with socio-economic-political settings constitutes an interdependent global environment and world ecology. By considering the physical-biological basis, the development should not exceed the ability of the environment to natural resources and services. The development is not just as an economic activity but as a process of qualitative and equitable growth. In process of development, sustainability should be positioned in political-institutional arrangements to restructure public power and create social decision-making. For social development, it is important to focus on the well-being of communities, creating jobs, and considering income distribution. By considering society as an interdependent and a world community, global economic growth cannot succeed with an uneven distribution of wealth. And it is worth remembering that environmental problems do not know territorial or economic barriers, national or international, and thus institutional arrangements of development will not just affect the quality of life of some nations. In the process of development, cultural values and beliefs also need to be recognized and considered time by time to guide and justify anthropic actions. Altogether these four key aspects direct the development toward sustainability.

Nature conservation is closely linked to sustainable development, particularly in the developing countries. It is evident that nature conservation is not possible without sustainable development which is aiming at both societal welfare and environmental protection [36–38]. In Vietnam as elsewhere of the developing world, protected areas do not receive the support of the people because they do not have a positive impact on people's livelihoods and do not support the development of cultural, social, political, natural, and human resources. People do not appreciate the management processes around the reserves [14]. And nature conservation only improves when relations between protected areas and communities are improved through the improvements of management processes and conservation and socioeconomic outcomes [38]. Therefore, in the 2030 Agenda for Sustainable Development, nature conservation is highly embedded across most of the Sustainable Development Goals [39]. Many protected areas around the world have already combined approaches to biodiversity conservation and sustainable development in which they facilitate sustainable economic activities in both aspects of environmental ecology and means of livelihood for communities [36, 40].

Additionally, convention on biological diversity (CBD) recognizes communities to play a huge role in biodiversity conservation and preserve traditional cultural values. Commitment to recognize and institutionalize community protected areas has been promoted globally and included in the regulation of the convention on indigenous peoples, local knowledge, and traditional resource use according to the customary law. In the context of CBD implementation, the contents of community protected areas are identified in Aichi Objective 11 on ensuring specific numbers of the area worldwide and Objective 18 with respect to knowledge, indigenous traditional initiatives, and practices in conservation and sustainable use of biodiversity. In order to implement the international agreements and treaties, many countries have promoted the institutionalization of protected areas managed and registered by communities in the global data system. The leading countries in this work include India, Nepal, China, Taiwan, Philippines, Malaysia, Indonesia, Thailand, Benin, Australia, Canada, Bolivia, and Madagascar, of which many neighbor Vietnam [41]. Therefore, this can be seen as opportunities for co-management to engage indigenous communities in arrangements, contributing to the cultural diversity and the emergence of sustainable societies across the world and so Vietnam alike.

2.2 Natural resource co-management

Co-management has been adopted internationally in response to the perceived failure of centralized management in natural resources [12, 33, 42–45]. It is a process of solving-problem management in which actors at different levels and scales interact to adjust their positions, roles, and activities to harmonize with emerging contexts and circumstances surrounding a natural resource [29]. In this process, power is a result, leading to modifications on the rule of the game and creating win-win solutions [46]. Today, it is defined as an arrangement where responsibility and right for resource management are shared between the government and user groups [44, 47], acknowledging the important role of the people who are living around the resources and impacting on resource uses and management [46]. Therefore, co-management arrangement often includes the devolution of responsibilities associated with day-to-day management of natural resources and in some cases a transfer of power and authority from national government agencies to communities and subnational governments [29, 48, 49]. In terms of participation, co-management arrangement engages local community groups or resource users in decision-making, implementation, and enforcement [50–52]. In order to ensure the participation, co-management focuses on developing effective local institutions and an enabling environment for sustainable management [37]. According to Jentoft, co-management may be the best available solution to the legitimacy problem because the center to the implementation of co-management is the design of new structure legitimized to bring together stakeholders for decision-making and implementation [53]. This design can be built up on existing arrangements at site levels or supported by donor funding and directed by central government in a top-down manner [54].

Based on the exercise of co-management at locals or on-site levels, adaptive governance evolved. It is a novel type of environmental governance that has arisen in systems characterized by large degrees of dynamism, complexity, and uncertainty [55, 56]. It combines learning, knowledge generation, and problem-solving of the adaptive management with the stakeholder power-sharing and conflict resolution of co-management [55, 57]. Folke et al. ([55], p. 8) broadly define adaptive co-management as "a process by which institutional arrangements and ecological knowledge are tested and revised in a dynamic, ongoing, self-organized process of trial-and-error," which is known to evolve through stages [58]. In context of conservation conflicts, adaptive co-management is evidenced supporting conflict solving by providing collaborative decision-making processes which involve all stakeholders equitably, trial innovative ideas, and include evaluation to provide learning [56, 59]. It also promotes local sustainability through capacity development and trust building, particularly as if protected area authorities become bridging organizations [57].

To combine nature conservation and sustainable development, many countries have shifted the modes of protected area management from centralized and non-participatory ones to co-management in order to benefit from co-management arrangement and exercise adaptive governance. In order to do so, it requires at least three factors, including the presence of institutional entrepreneurs, a dense central core of network actors, and the prevalence of horizontal ties and vertical linkages held by the community-based organizations responsible for the management of the resource [52]. Lawmakers can set up legislation for co-management in which it can shape decentralized management by recognizing and devolving responsibility to community-based management systems [60, 61]. And this is a necessary ingredient in co-management arrangement [53]. And in the case of Vietnam, although institutionalizing co-management in nature conservation is a must to achieve both ecological protection and sustainable development, it takes time to promote step by step to become legitimized.

3. Institutionalizing co-management in Vietnam protected areas for sustainable development: the case of Xuan Thuy National Park

To overcome deficiencies in the protected area management, Vietnam has piloted co-management in many protected areas since 2001 through a number of foreign-funded projects [62, 63]. To some extent, the pilots have not led to institutional reforms in protected area management arrangement, but the concept of co-management step-by-step has been included in official documents such as national strategy protected area management in 2003 because of its high potentials [20, 63]. However, in case of Xuan Thuy National Park, co-management has been applied and formed institutions for nature conservation and sustainable development. Based on in-depth interviews with the park management board and literature reviews, the case will be narrative in order to prevail the process of co-management installation in Xuan Thuy National Park and how it supports to overcome shortcomings emerged from the park's centralized management.

Xuan Thuy National Park was established in January 2003 and administered by the Nam Dinh DARD [64]. In 1988, 15 years prior to its establishment, the park became the first Ramsar site of Vietnam, and in October 2004 it was also acknowledged as the core zone of the Red River Delta Biosphere Reserve by UNESCO [65]. Before 2006, the institutions of Vietnam on natural conservation management strictly prohibited the use of natural resources in the core zone of protected areas. But at Xuan Thuy national park, in order to solve management problems to meet the requirements of local livelihood, the park has conducted interventions to implement policies to wisely use aquatic resources with the principle "Only allowing to exploit common aquatic species which are able to recover well, and absolutely prohibiting activities that lead to mangrove deforestation, depletion of natural resources, landscape changes and environmental pollution" .

The legal grounds for the interventions were not based on national regulations at that time but on Ramsar convention recommendations, advising wise and sustainable uses of wetland resources to meet local community needs for short-term benefits and for latterly long-term national and international benefits. On March 7, 2006, MARD issued an official document 511/BNN-KL expressing its agreement on permitting local community to exploit natural resource of mollusks (*Meretrix lusoria* and *Meretrix lyrata*) in Core Zone of Xuan Thuy National Park in conditions that DARD, who directly administrates the management board of Xuan Thuy National Park, was required to construct a feasible proposal on the exploitation management for MARD assessment before being ratified by Nam Dinh PPC. The feasible proposal had been finally completed after being discussed and consulted with state and local specialists. It was ratified by Nam Dinh PPC in the decree of *1951/QD-UBND dated August 24, 2006*. In the proposal, "applying co-management to sustainably use aquaculture resources in the area would harmonize integrated targets of nature conservation and development, simultaneously implementing Ramsar convention recommendations and International Biosphere Reserve criterions, and creating healthy environments for human and nature harmoniously living together" [64]. The effective implementation of the intervention was hoped to bring about sustainable development of local socio-economy because it would create incomes for local people, ensuring security and targets of natural resource protection. Local people became the main forces who would proactively and committedly carry out responsibilities toward nature conservation and sustainable development to maintain mutual benefits.

In this co-management arrangement, Giao Thuy DPC assigns agencies under its administration to collaborate with Xuan Thuy National Park to decide plans and methods for activities toward the mollusk exploitation. Nam Dinh PPC also delegates its authority to Giao Thuy DPC to decree the establishment of a management board of the aquaculture resource exploitation in Red River delta within Xuan Thuy National Park. The management board of the mollusk exploitation has been established and comprised of representatives from Giao Thuy DPC; Xuan Thuy National Park; the district divisions of fishery, natural resources and environment, finance planning, and taxes; Commune People's Committees of Giao An and Giao Thien; security forces of military and police, Giao Thuy District Station of Fishery Inspection; and Forest Protection Bureau of Xuan Thuy National Park. They are in charge of responsibilities relating to (1) planning the exploitation; (2) coordinating activities toward the exploitation; (3) collaborating with fishery branch to construct and manage the area sustainably; (4) checking, monitoring, and collaborating with functional agencies to ensure public security in the location; (5) implementing water surface allocations to households and collecting fees from the allocations and other contributions from households; (6) directing and supporting self-organization groups to implement their responsibilities in the exploitation area; (7) organizing dissemination and education activities toward Xuan Thuy National Park protection to communities; (8) enhancing scientific research and advanced technology exchanges to apply into sustainable exploitation; and (9) periodically reporting and monthly meeting to discuss plans, measurements, and responsibilities of self-organization groups.

Self-organization groups mentioned above are local people involved in the mollusk exploitation and selected by local people to represent them in the management board. The head of a group is a prestige person in a community, and the deputy head of a group is a commune security officer. Responsibilities of self-organization groups are also set up, including (1) protecting natural resource, environment, and security in the area; (2) checking activities in the area and its adjacent areas; (3) following the supervisions of the management board of the mollusk exploitation; (4) collaborating with the national park, military, police, and CPCs to implement their responsibilities; (5) mediating conflicts; (6) discovering and holding violations of regulations on nature resource management; (7) reporting violations to authorities to be measured; (8) weekly meeting to check activities and suggest coming activities; (9) weekly reporting to the management board; and (10) collecting information and reflecting aspirations of community to the management board or authorities to have suitable responses (**Figure 1**).

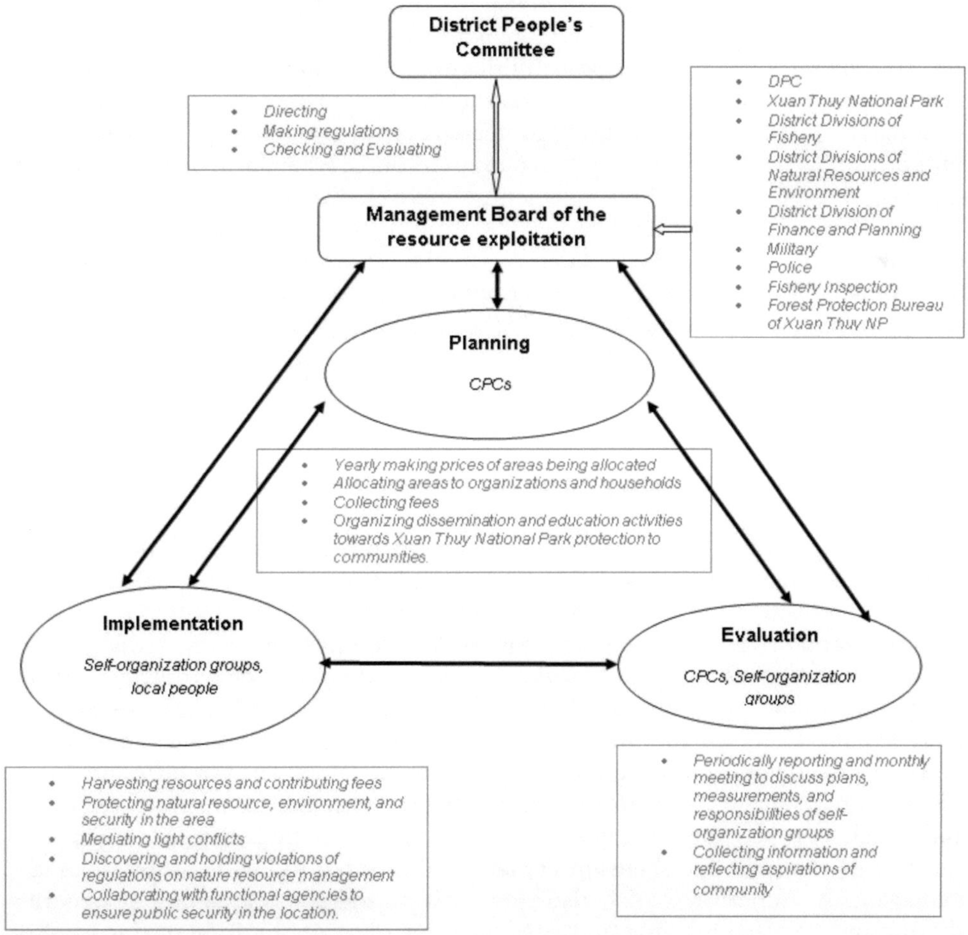

Figure 1.
Co-management arrangement in mollusk resource exploitation of Xuan Thuy National Park, Vietnam.

As a result, the park has received consensus agreement and supports of the local community. Resources targeted at nature conservation such as mangroves, birds, and wildlife together with ensuring the balance of aquatic resources and environmental protection in the area have been maintained. The local community has a stable income from the exploitation of aquatic resources. The average daily income from this activity is from VND 50 to 100 million. The income from extensive mollusk and shrimp farming models is about VND 200 billion/year. Sharing the above legitimate and reasonable benefits has enticed the local community to participate in nature conservation management of the park through many useful practical activities of the community, including the key mass organization sponsored by the park such as bird conservation club, mollusk farming association, community tourism management board, beekeeping club, and mushroom cooperatives [66]. This pilot project of co-management and the wise use of natural resources at Xuan Thuy National Park have been implemented from 2006 up to now.

The above scheme is a new breakthrough in the policy of protected area management in Vietnam. The relationships of related parties and benefit sharing are clearly institutionalized to secure and protect natural resources. After 4 years of pilot implementation, the local government has collected more than VND 2 billion from leasing land to exploiting natural mollusk seeds. Local communities also get tens of billions of revenue from legal exploitation of the resources while maintaining the quality of the environment . From that, the sustainable exploitation of natural seasonal mollusks resolved conflicts of illegal exploitation. Up to now, with clear and effective management institutions, there is an effective participation of stakeholders. The budget is tied to the responsibilities of local authorities and spent on supporting public welfare and nature conservation. This is a new direction to ensure sustainable financial mechanism for protected area management in Vietnam. It also supports to overcome the shortages in human resource when local people are really engaged in resource use and management. Moreover, after the intervention of Xuan Thuy National Park, there has been a compatible policy shift at national level. The Decision No. 186/2006/QD-TTg dated August 14, 2006, of the prime minister on the promulgation of forest management regulations has recognized the wise use of natural resources in protected areas. This is an innovative approach in which some common species are allowed for harvesting and supporting community livelihood development and nature conservation of protected areas [66].

4. Initiatives in legitimizing co-management of Vietnam protected area

According to KimDung et al. [67], Vietnamese implementation of co-management in protected areas is best defined as "administrative," reflecting the stronger role of the central state over communities and others. The authors found that the existing Vietnam policy and legislation provides a foundation for the development of co-management through diversifying the type and number of actors involved in protected area management, matching with the expectations of co-management arrangement. However, the "administrative co-management" arrangement remains centralized and not yet based on the mobilization of actors' self-interests and economic motivation. The state maintains rights or ownership over protected areas and remains unclear about the notion of "community" in the policy, leading to constraining the practice of customary laws and community-based protected area management. Moreover, there is also a lack of legal and policy guidance on benefit sharing and reinvestment into protected area conservation, limiting the incorporation of economic actors in nature protected areas [67, 68].

As argued by KimDung et al., any amendments to co-management in Vietnam would remain in the hands of lawmakers, who maintain control over law enforcement and capacity building, mobilizing outside support, and creating mechanisms for information transparency [68]. The unique practice of Vietnamese co-management arrangement has been able to implement in Xuan Thuy National Park because it gained strong political supports from global to national and local levels (MARD, PPC, DPC, and CPCs surrounding the park), consequently harmonizing conflicts between the dynamics of local livelihoods, market demands, and nature conservation. In this model, roles of state forest rangers become blurred leading to a question whether they are the force in need as if the engagement of local people is provided. If policy modifications on protected area management are not taking place and innovated, conflicts on natural resources will increase, and the government might have lost their roles to control protected area resource in the context of economic dynamics in which demands on natural resource are highly increased, while the resource becomes short and rare. Recently, the 2017 Forest Law has marked an important milestone in recognizing communities as one of the seven types of forest owners (Article 8). The sacred forest of community is classified as a landscape protection forest under the protected forest system. The Law on Fisheries 2017 also recognizes the co-management model between state forest owners and related communities. In the coming time, the recognition of community conservation areas is advocated by an NGO, People and Nature Reconciliation (PanNature), to be considered and included in the revised Biodiversity Law, linked to the content of access to genetic resources, benefit sharing, and indigenous knowledge development in biodiversity conservation [41].

Additionally, both government agencies and NGOs can be bridging organizations who attempt to install co-management into protected areas. In the case of Xuan Thuy National Park, the strong vertical support by the MARD, in the context of national legislation related to Ramsar, has allowed the management board to largely bypass the control of the PPC administration to not only solve conflicts over resource exploitation but also provide learning and introduce co-management of protected area in Vietnam. To some extent, it is regarded as an institutional entrepreneur for the very first co-management of natural resources and sustainable livelihood development in protected areas of Vietnam. It recognizes the mutual benefit and interdependencies between local people, the park management board, and commune governments, providing the connection between the actors at different levels and the cooperation in practical real-life arrangement to solve resource problems [69–71].

Moreover, in order to support the development of co-management, NGOs have developed capacities as bridging organizations between protected areas, communities, and government by coordinating collaboration across levels, sectors, and knowledge systems. People and Nature Reconciliation, the Centre for Marine Life Conservation and Community Development (MCD), Vietnam National Park and Protected Area Association (VNPPA), and Fauna and Flora International (FFI), among the NGOs working in Vietnam nature conservation and development, have attempted to foster information exchange and create a common vision of co-management across multiple levels [72]. Those are positive initiatives for co-management developed and brought benefits to communities of protected areas in Vietnam. Finally, to achieve both sustainable development and nature conservation, Vietnam keeps the process of institutionalizing co-management. Next steps should be focused on long-term agreements to ensure the rights to access and share benefits and practices of sacred forest protected areas. Opportunities for economizing the management of protected area and creating sustainable development mechanisms need to be prioritized through the scheme of co-management.

5. Conclusion

Co-management provides opportunities for the integration of nature conservation and sustainable development. Therefore, it is necessary to institutionalize this mechanism in the context of developing countries like Vietnam. In order to do so, the Vietnamese co-management arrangement needs more flexibility and adaptability to adjust actors' positions and roles, promoting more policy modifications in protected area management to harmonize practices in and around the areas. One of the most important recommendations is legitimizing the practices of using natural resources in a certain extent at protected areas. By doing so, the role of local people is adjusted to become resource users and responsible for the sustainable use of the resources while keeping their eyes on the other resources for nature conservation in the area of protected areas. The unique practice of this co-management arrangement in Xuan Thuy national park harmonizes conflicts between the dynamics of local livelihoods, market demands, and nature conservation. Although there are some initiatives in this long-run process, potentials reveal that it is worth to pursue for a better sustainable future in which communities truly benefit from protected areas. To do so, the government as lawmaker should lead to facilitate the process in line with the support from communities and civil society such as a force of NGOs.

Conflict of interest

The author declares that there is no conflict of interests in this work.

Author details

Nguyen Kim Dung
University of Science, Vietnam National University, Ho Chi Minh City, Vietnam

*Address all correspondence to: kimdunguyen@gmail.com; ntkdung@hcmus.edu.vn

References

[1] Birdlife International. Vietnam IBA Status [Online]. 2016. Available from: http://www.birdlife.org/datazone/country/vietnam/ibas

[2] Pilgrim JD, Tu ND. Background Paper on Threatened and Alien Species in Vietnam and Recommendations for the Content of the Biodiversity Law. Birdlife International Program; Hanoi, Vietnam; 2007

[3] USAID. Vietnam Tropical Forest and Biodiversity Assessment. Sun Mountain International and The Cadmus Group, Inc.; 2013

[4] MONRE. 2010 Vietnam Environmental Report: Overview of Vietnam Environment. Hanoi: Ministry of Natural Resource and Environment; 2011

[5] PanNature. Quarterly Policy Review. Hanoi: People and Nature Reconciliation; 2011

[6] Sunderlin W. Poverty alleviation through community forestry in Cambodia, Laos, and Vietnam: An assessment of the potential. Forest Policy and Economics. 2006;**8**(4):386-396

[7] To XP. Why did the forest conservation policy fail in the Vietnamese uplands? Forest conflicts in Ba vi National Park in northern region. International Journal of Environmental Studies. 2009;**66**(1):59-68

[8] McElwee P. Resource use among rural agricultural households near protected areas in Vietnam: The social cost of conservation and implications for enforcement. Environmental Management. 2010;**45**:113-131

[9] Sikor T, Phuc T. Illegal logging in Vietnam: Lam tac (forest hijackers) in practice and talk. Society & Natural Resources. 2011;**24**(7):688-701

[10] Cuong T, Dung N, Trang T, Minh P, Le P. Some Contents of Vietnamese Biodiversity Management. Hanoi: Department of Biodiversity Conservation; 2009

[11] McElwee P. Forest Are Gold: Trees, People, and Environmental Rule in Vietnam. Washington: University of Washington Press; 2016

[12] KimDung N, Bush S, Mol A. Administrative co-management: The case of special use forest conservation in Vietnam. Environmental Management. 2013;**51**(3):616-630

[13] Khai HV, Yabe M. The demand of urban residents for the biodiversity conservation in U Minh Thuong National Park, Vietnam. Agricultural and Food Economics. 2014;**2**(1):10

[14] Vietnam Government. Decision of the Prime Minister to Approve the Management Strategy for a Protected Area System in Vietnam to 2010; Hanoi; 2003

[15] Development Partnerships. Report on Vietnam Development 2010: Natural Resource Management. Lebros, Hanoi, Vietnam; 2010

[16] VEPA. Overview of Wetlands Status in Vietnam Following 15 Years of Ramsar Convention Implementation. Hanoi, Vietnam: Vietnam Environment Protection Agency (VEPA), The World Conservation (IUCN Vietnam) Mekong Wetlands Biodiversity Conservation and Sustainable; 2005

[17] PARC Project. Creating Protected Areas for Resource Conservation Using Landscape Ecology (PARC) Project VIE/95/G31&031. Policy Brief: Building Vietnam's National Protected Areas System—Policy and Institutional Innovations Required for Progress. Hanoi, Vietnam: Vietnam Government; 2006

[18] Vietnam Government. Decision 08/2001/QĐ-TTg Dated 11/01/2001 of Prime Minister on Issuing Regulations on Management of Special-Use Forests, Environmental Protection Forests and Natural Production Forests. Hanoi, Vietnam: Vietnam Government; 2001

[19] Swan S. Co-management: Concepts and practices in Vietnam. In: Proceedings of The National Workshop on Co-Management Concept and Practice in Vietnam. Soc Trang, Vietnam; 2010

[20] ICEM. Vietnam National Report on Protected Areas and Development. Indooroopilly, Queensland, Australia; 2003. p. 66

[21] KimDung N. Administrative Co-Management in Special Use Forests of Vietnam. Wageningen, The Netherlands: Wageningen University; 2017

[22] Sunderlin WD, Ba HT. Poverty Alleviation and Forests in Vietnam. Indonesia: CIFOR; 2005

[23] Government of Vietnam. The Fifth National Report on the Implementation of Convention on Biological Diversity. Hanoi, Vietnam: MONRE; 2014

[24] PanNature. Handbook: Skills for Enhancing Community Participation in Forest Management. Hanoi: Hong Duc Publishing House; 2014

[25] Zingerli C. Colliding understandings of biodiversity conservation in Vietnam: Global claims, national interests, and local struggles. Society & Natural Resources. 2005;**18**(8):733-747

[26] Boissiere M, Sheil D, Basuki MW, H L. Can engaging local people's interests reduce forest degradation in Central Vietnam? Biodiversity and Conservation. 2009;**18**(10):2743-2757

[27] McElwee P. You say illegal, I say legal: the relationship between illegal' logging and land tenure, poverty, and forest use rights in Vietnam. Journal of Sustainable Forestry. 2004;**19**(1/2/3):97-135

[28] Carlsson L, Berkes F. Co-management: Concepts and methodological implications. Journal of Environmental Management. 2005;**75**:65-76

[29] Armitage DR, Plummer R, Berkes F, Arthur RI, Charles AT, Davidson-Hunt IJ, et al. Adaptive co-management for social-ecological complexity. Frontiers in Ecology and the Environment. 2009;**7**(2):95-102

[30] Bown NK, Gray TS, Stead SM. Co-management and adaptive co-management: Two modes of governance in a Honduran marine protected area. Marine Policy. 2013;**39**:128-134

[31] Noble BF. Institutional criteria for co-management. Marine Policy. 2000;**24**:69-77

[32] Pomeroy RS. Community-based and co-management institutions for sustainable coastal fisheries management in Southeast Asia. Ocean and Coastal Management. 1995;**27**(3):143-162

[33] McBeath J, Rosenberg J. Advances in Global Change Research: Comparative Environmental Politics. The Netherlands: Springer; 2006

[34] Egelston AE. Sustainable Development: A History. London: Springer; 2013

[35] Rieckmann M, AdomBent M, Hardtle W, Aguirre P. Sustainable development and conservation of biodiversity hotspots in Latin America: The case of Ecuador. In: Biodiversity Hotspots Distribution and Protection of Conservation Priority Areas. The Netherlands: Springer; 2011

[36] Rashid AZMM, Khan NA. Role of co-management organizations in protected area governance: Some observations from the Chunati wildlife sanctuary. In: Forest Conservation in Protected Areas of Bangladesh. Switzerland: Springer International Publishing; 2014

[37] Bennett NJ, Dearden P. Why local people do not support conservation: Communities perceptions of marine protected area livelihood impacts, governance and management in Thailand. Marine Policy. 2014;**44**:107-116

[38] UNECE. Biodiversity at the Heart of Sustainable Development, Regional Forum on Sustainable Development for the UNECE Region. Geneva; 2018

[39] Gammanpila M, Jayantha M, Wijeyaratne S, Amarasinghe US. The dwindling community-based management strategies in the brush park fishery of a tropical estuary: Need for co-management. Ocean and Coastal Management. 2019;**167**:145-157

[40] PanNature. Policy Discussion: Biodiversity Protection and Cultural Preservation Through the Construction of Community-Based Protected Area System2018. p. 16

[41] Berkes F, George P, Preston R. Co-management: The evolution of the theory and practice of joint administration of living resources. In: Common Property Conference, The Second Biennial Conference of the International Association for the Study of Common Property; Winnipeg, Manitoba. 1991

[42] McCay B, Svein J. From the bottom up: Participatory issues in fisheries management. Society and Natural Resources. 1996;**9**:237-250

[43] Sen S, Nielsen J. Fisheries co-management: A comparative analysis. Marine Policy. 1996;**20**:405-418

[44] Arthur R. Developing, Implementing and Evaluating Policies to Support Fisheries Co-Management. MRAG; 2005

[45] Sandström C. Institutional dimensions of co-management: Participation, power, and process. Society and Natural Resources. 2009;**22**(3):230-244

[46] Plummer R, Fennell D. Managing protected area for sustainable tourism: Prospects for adaptive co-management. Journal of Sustainable Tourism. 2009;**17**(2):149-168

[47] Pomeroy RS, McConney P, Mahon R. Comparative analysis of coastal resource co-management in the Caribbean. Ocean and Coastal Management. 2004;**47**:429-447

[48] Cinner JE, Daw TM, McClanahan TR, Muthiga N, Abunge C, Hamed S, et al. Transitions toward co-management: The process of marine resource management devolution in three east African countries. Global Environmental Change. 2012;**22**:651-658

[49] Jentoft S. User participation in fisheries management. Lessons drawn from international experiences. Marine Policy. 1995;**19**:227-246

[50] Berkes F. Devolution of environment and resources governance: Trends and future. Environmental Conservation. 2010;**37**(4):489-500

[51] Alexander SM, Armitage D, Charles A. Social networks and transitions to co-management in Jamaican marine reserves and small-scale fisheries. Global Environmental Change. 2015;**35**:213-225

[52] Jentoft S. Legitimacy and disappointment in fisheries management. Marine Policy. 2000;**24**:141-148

[53] Nunan F, Hara M, Onyango P. Institutions and co-management in east African inland and Malawi fisheries: A critical perspective. World Development. 2015;**70**:203-214

[54] Folke C, Hahn T, Olsson P, Norberg J. Adaptive governance of social-ecological systems. Annual Review of Environment and Resources. 2005;**30**:441-473

[55] Butler J, Young J, McMyn I, Leyshon B, Graham I, Walker I, et al. Evaluating adaptive co-management as conservation conflict resolution: Learning from seals and salmon. Journal of Environmental Management. 2015;**160**:212-225

[56] Fabricius C, Currie B. Adaptive co-management. In: Adaptive Management of Social-Ecological Systems. Dordrecht: Springer; 2015. pp. 147-179

[57] Plummer R, Baird J. Adaptive co-management for climate change adaptation: Considerations for the barents region. Sustainability. 2013;**5**(2):629-642

[58] Silva P, Cabral H, Rangel M, Pereira J, Pita C. Ready for co-management? Portuguese artisanal octopus fishers' preferences for management and knowledge about the resource. Marine Policy. 2019;**101**:268-275

[59] Fisher R. Decentralization and Devolution in Forest Management: A Conceptual Overview. Decentralization and Devolution of Forest Management in Asia and the Pacific. RECOFTC Report; 2000

[60] Dang T, Visseren-Hamakers I, Arts B. The institutional capacity for forest devolution: The case of forest land allocation in Vietnam. Development and Policy Review. 2017;**35**(6):723-744

[61] Watt AB, Quan LT. Evaluation Report: Sustainable Livelihoods and Participatory Wetlands Conservation in the Plain of Reeds. Atlanta, GA, USA: CARE; 2007

[62] Sandbrook C, Trong NP, Lee K, Grieg-Gran M. Co-Management of Forests and Wildlife in the Bi Doup-Nui Ba (BD-NB) Nature Reserve; Vietnam [Online]. 2010. Available from: http://www.iied.org/sustainable-markets/keyissues/environmental-economics/co-managementforests-and-wildlife-vietnam [Accessed: December 28, 2011]

[63] N.D. PPPC. A Piloting Project on Sustainable Use of Meretrix Lusoria and Meretrix Lyrata Resources in a Wetland of Red River Estuary of Xuan Thuy National Park. Nam Dinh Province, Vietnam: People's Committee Council of Nam Dinh Province; 2006

[64] Hong P, Cach N. Xuan Thuy National Park: Management. Hanoi: VEPA_MCD; 2007

[65] Nguyen Viet Cach. Economizing biodiversity conservation—The typical case of Xuan Thuy National Park (Kinh tế hóa công tác bảo tồn đa dạng sinh học—Trường hợp điển hình tại vườn quốc gia Xuân Thủy). Journal of Environment. 2010. http://tapchimoitruong.vn/pages/article.aspx?item=Kinh-t%E1%BA%BF-h%C3%B3a-c%C3%B4ng-t%C3%A1c-b%E1%BA%A3o-t%E1%BB%93n-%C4%91a-d%E1%BA%A1ng-sinh-h%E1%BB%8Dc---Tr%C6%B0%E1%BB%9Dng-h%E1%BB-%A3p-%C4%91i%E1%BB%83n-h%C3%ACnh-t%E1%BA%A1i-v%C6%B0%E1%BB%9Dn-qu%E1%BB%91c-gia-Xu%C3%A2n-Th%E1%BB%A7y-40727

[66] KimDung N, Bush SR, Mol AP. The Vietnamese state and administrative co-management of nature reserves. Sustainability. 2016;**8**(3):292

[67] KimDung N, Bush SR, Mol AP. The Vietnamese legal and policy framework

for co-management in special-use forests. Forests. 2017;**8**:262

[68] Plummer R, FitzGibbon J. Co-management of natural resources: A proposed framework. Environmental Management. 2004;**33**(6):876-885

[69] Armitage D, Berkes F, Doubleday N. Adaptive Co-Management: Collaboration, Learning, and Multi-Level Governance. Vancouver, BC, Canada; 2010

[70] Ostrom E, Cox M. Moving beyond panaceas: A multi-tier diagnostic approach for social-ecological analysis. Environmental Conservation. 2010;**37**(4):451-463

[71] KimDung N, Bush SR, Mol AP. NGOs as bridging organisations in managing nature protection in Vietnam. Journal of Environment & Development. 2016;**25**(2):191-218

[72] Government of Vietnam. Decree 156/2018/NĐ-CP Dated 16/11/2018 on Regulations of Detailed Implementation of Forestry Law. Hanoi, Vietnam: Government of Vietnam; 2018

Chapter 5

On the Road to Sustainability? A Review of a Half-Century of Biodiversity Conservation Successes in Nepal and Some Thoughts on Future Needs

Joel T. Heinen, Nabin Baral, Prakash K. Paudel and Jay P. Sah

Abstract

We review the history, over the past half-century, of biodiversity conservation legislation and programs in Nepal. We especially consider how they have evolved in light of some earlier concerns resulting from, for example, the strict "fines and fences" conservation approaches first implemented in the 1970s, to pressing issues that emerged over time such as park-people and wildlife-human conflicts, poaching and illegal wildlife trade. We also consider how the implementation of international conservation agreements and demographic and political changes have affected conservation programs in Nepal. We finish by discussing conservation in the context of sustainable development and conclude with some thoughts on future research and managerial needs in a rapidly-changing world.

Keywords: biodiversity, conservation, Nepal, protected areas, sustainable development

1. Introduction

Nepal is a poor land-locked, mountainous country of 147,181 sq. km in area and approximately 30 million residents. Mountainous regions in general, given their climatic diversity, tend to be high in species diversity and endemism (e.g. [1]). Nepal is located at the boundary of the Palearctic and Indo-Malayan zoogeographic provinces at relatively low subtropical latitudes with elevations ranging from ca. 100 m above mean sea level (asl) to the world's highest peaks. For those reasons, it boasts very high species diversity, many centers of adaptive radiation and high degrees of endemism and animals [2–6]. Nepal is recognized as being very important for biodiversity within the Asia-Pacific region [7, 8] and it boasts several Global 200 Ecoregions, two endemic bird areas, and it's a global biodiversity hotspot (e.g. [9, 10]).

Here we explore the history of conservation programs in Nepal and how they have expanded over the past half century with the objective of asking how sustainable such programs are likely to be into the future. Our review relies on a detailed literature review, interviews over many years with government officers and fellow academics and, collectedly, among us, over 80 years of studying biodiversity and conservation programs in the country. We focus on many conservation programs that have shown promise (e.g. [11–14]) and we consider some taxon-specific conservation successes and concerns (e.g. [15, 16]). We pose this review in the context of sustaining biodiversity in an uncertain world given rapid climate change and its likely affects (e.g. [17, 18]) and we finish with some thoughts about future research and management needs.

2. Meager beginnings

Much of the early history of conservation programs in Nepal has been covered in depth elsewhere (e.g. [19]). Here we give a brief background. The country was largely closed to the rest of the world until the 1950s; it was still mostly forested and underdeveloped at that time and had large wildlife populations and extensive habitat. Among the first major international projects, beginning in the 1950s, was that of the United States Agency for International Development (USAID), in conjunction with the World Health Organization (WHO), to eradicate the virulent strain of malaria endemic to the lowland (terai) region along Nepal's southern border with India. With malaria eradication, the government began moving large numbers of land-poor mountain farmers to the terai and rapid rates of deforestation and poaching ensued.

With the events described above, the successful ascent of Sagarmatha (Mr. Everest) in 1957 [20], and growing numbers of westerners visiting in the 1960s, it was apparent that the Nepal had potential as a major tourist destination and that rampant deforestation and poaching in the 1950s and 1960s were not sustainable [21, 22]. Economic development and conservation were both major concerns of King Mahendra during this time and large areas of the terai had started to undergo faunal collapse of large mammals due to these pressures [23]. In response, the king formed the rhino patrol in the central terai, in what is now part of Chitwan National Park [24]. The government then requested the first of several large projects over the years to study and promote conservation beginning in the late 1960s under the United Nations Development Program (UNDP). That first project identified important wildlife populations, proposed the first areas to be protected (some of which had been royal hunting reserves) and made many recommendations to the government about necessary legislation and staffing for conservation.

2.1 The seventies: modern conservation implemented

Based on recommendations from that UN project, by 1973, the Department of National Parks and Wildlife Conservation (DNPWC) was formed (from an office within the Forest Department) as a separate entity within the Ministry of Forests and Soil Conservation, and the first protected area (PA; Chitwan National Park) was formed [25–27]. That year also saw the passage of the National Parks and Wildlife Conservation Act (the Act) that recognized four types of protected areas (PAs): national parks (NPs), wildlife reserves (WRs), strict nature reserves (SNRs) and hunting reserves (HRs). The first three correspond to International Union for the Conservation of Nature (IUCN) Protected Area Categories II, IV and Ia, respectively, while HRs do not meet IUCN standards [28]. No separate SNRs were designated because the DNPWC decided they would be impractical to enforce.

The act also published a list of protected species and hunting regulations for others. Many Nepali nationals were also sent abroad during the 1970s and 1980s, under a number of bi- and multi-lateral international projects, to pursue professional degrees in conservation. Growing numbers of foreign nationals also pursued wildlife research in Nepal from those years to the present. This was greatly helped by the Smithsonian Institution's funding of Nepal's first permanent field site in Chitwan NP to study tigers and their prey [21, 29, 30]. That facility still exists and is now managed by the Nepal Trust for Nature Conservation (below).

Along with many other developing countries [31–35], Nepal at first adopted a strict "fences and fines" conservation model and, by the end of the 1970s, it was apparent that this led to various types of park-people conflicts [36, 37]. While living in the vicinity of PAs can have measureable economic benefits (e.g. [38]) crop loss [39] is common around PAs as wildlife populations recover, and core PAs in Nepal were off limits to activities, such as fodder and thatch grass collection, upon which rural villagers had previously depended [40, 41]. Attacks by wildlife on humans, including fatalities, have also been of growing concern from the 1970s to the present [42–44].

In one particularly disastrous decision, villagers whose families had lived in the area for centuries were removed from Rara NP (ca. 3000 m asl) to the western terai, and many succumbed to malaria (e.g. [45]). With these events, the DNPWC began changing some rules by, for example, allowing private inholdings in the other Himalayan PAs such as Sagarmatha and Langtang NPs, and allowing thatch grass removal for periods of time during the dry season in the terai PAs (e.g. [46, 47]). The former policy also had the advantage of allowing private homeowners to rent rooms to trekkers for the growing tourist markets and thus (presumably) to allow locals direct economic benefits from tourism [48–51].

2.2 Expanding and experimenting: the eighties and nineties

The 1980s saw immense growth in the conservation sector of Nepal [19]. Several major international non-governmental organizations (INGOs), as well many of national-level NGOs, in support of conservation efforts began during this period and into the 1990s, and the PA system approximately doubled in area. INGOs such as the World Wildlife Fund (WWF), The Mountain Institute (TMI), the International Union for the Conservation of Nature (IUCN) and the Integrated Center for International Mountain Development (ICIMOD) all opened offices in Kathmandu during the decade of the 1980s. ICIMOD is a regional organization that includes eight Himalayan countries as members. It was founded under the UN's Education, Scientific and Cultural Organization (UNESCO) Man and Biosphere Program. The Nepal Trust for Nature Conservation, (NTNC) also began during this period. They all have since expanded greatly both in scope and funding within Nepal.

Many other national and local NGOs opened in the 1980s and 1990s as a result of democratization and attempts to decentralize power [52]. Civil society in all sectors expanded greatly during this period and the phenomenon shed light on many pressing issues. For example, the NGO Wildlife Conservation Nepal (WCN) focuses mostly on illegal wildlife trade and has been instrumental in breaking up several criminal organizations [53]. Several local NGOs focus on researching the potential for cultivating and marketing medicinal and aromatic plants (MAPS; [54]), a very important economic activity in the region (e.g., [55, 56]), and still others have formed around, for example, bird conservation (e.g. [57]). These efforts have had the effect of greatly increasing awareness of conservation in Nepal and attracting more funding for conservation efforts at multiple scales.

It was also during this period that Nepal became Party to the Conventional on International Trade in Endangered Species of Wild Fauna and Florida (CITES), the convention on Wetlands of International Importance (Ramsar; [58]) and the World Heritage Convention (WHC), all of which positioned the country at the forefront of global conservation [59, 60]. Koshi Tappu WR became Nepal's first wetland of international importance under the Ramsar convention [47]; the country has since listed seven more sites (e.g. [58]) and developed a wetlands policy to promote sustainable uses, in part due to the high ecosystem values associated with wetlands [61]. Sagarmatha and Chitwan NPs are world heritage natural sites and Nepal has many other potential WHC natural sites not yet listed including several of its other Himalayan PAs. Nepal is rich in WHC cultural sites such as the ancient temple complexes in Kathmandu Valley and Lumbini, the birthplace of the founder of Buddhism. The WHC can be especially important in promoting natural and cultural tourism, and the tourism sector has grown greatly over the years within Nepal (e.g. [62–64]).

The 1980s and 1990s (to the present) saw the passage of legislation within the PA and forestry sectors to allow for greater decentralization and democratization of natural resource use and governance, thus empowering local communities. Amendments were passed that allowed for the development of conservation areas (CAs) as a new type of PA recognized in law, as well as buffer zones (BZs) around NPs and WRs [19]. CAs are not as strictly protected as NPs or WRs, but are set aside with the intention of providing some additional protections for biodiversity, while empowering local communities to make decisions about sustainable use of local resources. Nepal's CAs correspond to IUCN Category VI [28] and much of recent expansion of the PA system has been through designating CAs in the Himalayas (**Table 1**).

In many ways, the rights and general rules granted under CA and BZ legislation reflected the earlier movement toward community forestry first implemented by the forest department and Nepal-Australia forestry project in a few districts in the mountains, which then spread to the rest of the country (e.g. [65]). They allow for some degree of local community ownership via permitting user groups to exclude outsiders and decide harvest regimes for local needs [66]. The BZ and CA laws also allow for revenues generated from tourist entry fees to be used by local communities for development projects. While government agencies maintain the power to approve plans, the regulations go a long way toward community-based conservation. They also require female and minority representation on user committees but that has proven difficult to attain in the highly socially- and gender-stratified society of Nepal (e.g. [67–69]).

The 1990s also saw the beginnings of transboundary conservation efforts between Nepal and India and Nepal and China (e.g. [19]). To date, a series of bi- and multilateral meetings have taken place in Kathmandu on these issues despite China and India's mutual hostilities over their international borders, and a number of reserves have been established in both countries that abut or come close to Nepali PAs. They include the massive Qomalanga Reserve in Tibet that borders six of Nepal Himalayan PAs, to Indian reserves that border several of Nepal's terai NPs and form additional habitat and connectivity for wildlife populations [70]. While co-management has not been achieved through these efforts, the three countries recognize that nature does not conform to political borders and coordination of efforts is needed.

2.3 Into the twenty-first century

By the mid 1990s, Nepal became party to the convention on biological diversity and spent several years developing a biodiversity action plan [71]. This commitment partly led to the development of national policy and, later, legislation for the study and domestic propagation of medicinal and aromatic plants (MAPs) as part of its non-timber forest products policy (NTFPs, [54]). Value addition could be of great potential importance because Nepal has been a supplier of raw materials to the Ayurvedic medicinal industry based in India for centuries. Via CBD and the NTFP policies under implementation, the country is now positioning itself to be

Name	Date	Area	BZ area
a. National parks			
Chitwan NP	1973	932	360
Sagarmatha (Mt. Everest) NP	1976	1148	443
Langtang NP	1976	1710	660
Rara NP	1976	106	41
Shivapuri NP*	1976	159	61
Suklaphanta NP*	1976	305	118
Khaptad NP	1984	225	87
Parsa NP*	1984	637	246
Shey-Phoksundo NP	1984	3555	1373
Makalu-Barun NP	1992	1500	580
Banke NP	2010	550	210
b. Conservation areas			
Annapurna CA	1985	7629	
Blackbuck CA	1997	16	
Kanchenjunga CA	1997	2035	
Manaslu CA	1998	1642	
Api Nampa CA	2010	1903	
Gaurishankar CA	2010	2179	
c. Reserves			
Koshi Tappu WR	1976	175	68
Dhorpatan HR	1987	1325	
Totals		27,731	4247
Grand total		31,978	

The abbreviations are: NP, national parks; CA, conservation area; HR, hunting reserve and WR, wildlife reserve. Most of the older parks have been increased in area and only current (2019) areas are given. The core areas of all PAs include about 19% of the land area of Nepal. With buffer zones (BZ), the figure increases to about 22% of Nepal. Note that some NPs (*) were originally designated as WRs and later upgraded.

Table 1.
The names, establishment dates and areas (sq. km) of PAs of Nepal, with their buffer zone areas where appropriate.

a producer of some medicines and a domestic cultivator of some species of MAPS to produce greater supplies for growing markets and to improve rural livelihoods through biodiversity conservation and use, a major goal of CBD. This has shown great promise for conservation of rare MAPs in other regions as well (e.g. [72]).

These efforts show great promise and have garnered a good deal of NGO interest and support. Although CITES (above) has been difficult to implement for many reasons worldwide (e.g. [73]), and Nepal has faced some difficult wildlife poaching and smuggling issues for several decades (e.g. [74, 75]), the passage of comprehensive national legislation to implement the convention has proven important in raising awareness of wildlife trafficking and shows some potential for reducing poaching threats [53]. Both the NTFP and wildlife trade policies are now national laws that attempt to address these pressing issues. Both require the formulation and regular meetings of national-level cross-sectoral advisory groups that include high government officials to oversee implementation.

Political instability is common in developing countries (e.g. [76, 77]) and, from 1996 to 2006, the Maoist insurgency was impacting all aspects of society in Nepal including the conservation sector [78]. Despite major setbacks, the government continued to implement conservation conventions via expanding WHC and Ramsar designations within the country, developing sustainable use policies for BZs and CAs, and drafting the wildlife trade, wetland and NTFP policies. Regulations were also adopted for better PA management planning [79]. These all had the effect of furthering sustainability locally and further advertising Nepal's vast cultural and natural heritage to its growing tourist sector. These efforts made visiting Nepal more commonplace after several years of decreased tourist arrivals during the insurgency. They also improved financial resources for conservation via tourist entries into PAs and foreign projects to promote those policies.

Both WHC and Ramsar (above) maintain trust funds to help developing countries with projects in furtherance of those conventions. They thus present opportunities but also imposed costs in the form of the need to develop national legislation and policies to further those efforts. Throughout the prior period, Nepal relied on its international treaty legislation to implement any number of global agreements to which it was party, and it took at least another decade or more for the country to develop national implementing or enabling legislation [53]. The result was poor compliance. For example, The DNPWC was designated the management authority for CITES, but had no jurisdiction outside of PAs. The department of forests was the secondary management authority for most of the country (i.e. outside PAs) but had no law enforcement unit to implement wildlife legislation. While Ramsar is perhaps easier to implement, the lack of enabling legislation, and the Maoist insurgency, postponed adding more Nepali sites to the international list for years. These issues have, thankfully, been addressed through the drafting of national enabling legislation, but much research remains to be done in terms of improved transparency and efficiency of implementing agencies and laws.

2.4 More recently...

Nepal began its first wildlife translocations in the 1980s by moving rhinos from the large population in Chitwan NP to Bardia NP, where the species had been extirpated [24]. This program was successful in that the Bardia population expanded well into the 1990s, and more animals were moved in that decade as well. The Chitwan population also continued to grow. But the Maoist insurgency had the effect of loosening law enforcement within PAs and many animals in both NPs were poached. Both populations have been recovering since.

Given the initial success of the rhino translocation, the fact that the insurgency is over, and the fact that all of Nepal's terai reserves have undergone some degree of faunal collapse of large mammal species [23], more translocations have been planned or attempted. Wild buffalo were proposed for reintroduction many times previously (e.g. [16]) and funds were garnered from the United States government for this purpose in 2014 [80]. Buffalo were moved from the large population in Koshi Tappu WR to Chitwan NP in 2017, as were swamp deer from Suklaphanta NP to Chitwan NP. To date, the buffalo translocation appears to have been a success but the swamp deer translocation does not. In the latter case, most of the translocated animals have died. Research is ongoing into these attempts and large populations of buffalo in Koshi Tappu and swamp deer in Suklaphanta remain, so more projects could be carried out if the situations warrant such.

By the early 2000s, it was well known that snow leopards had recolonized Sagarmatha NP after a long absence [81]. This, and several other observations in Himalayan PAs (e.g. records of brown bears in Manaslu CA and argali sheep in Annapurna CA), led some researchers to hypothesize that the Maoist insurgency, negative as it was for terai wildlife, may have been perversely positive for Himalayan wildlife. The persistence of large mammals world-wide is considered to be a measure of ecosystem health (e.g. [82]) and, during the insurgency, many mountain inhabitants sold livestock and moved elsewhere, especially to Kathmandu valley and other urban areas, to escape its effects. The dearth of tourism for several consecutive years in the Himalayas may also have had the effect of reducing many other types of disturbances at high elevations (e.g. [83]). For these reasons, several researchers have explored possibilities for reintroductions of mountain ungulates in places where they have been extirpated as well (e.g. [84]).

Lastly, environmental and ecological economics began as a academic disciplines in the 1980s and have greatly expanded. They are important for policy makers to understand the values of biodiversity at species, community and ecosystem levels in national economies. Research in Nepal has ranged from using contingent valuation [85] to assess the value of vultures [68, 69], using travel cost methods to explore the values of wetlands [86], to assessing possibilities for, and economics of, implementing REDD (reducing emissions from deforestation and degradation, [87]). Various findings have shown that fees foreign tourists are willing to pay for entry permits into PAs are frequently higher than what governments charge (e.g. [88, 89]). Differential fees are the norm for residents versus foreign nationals in many places, but such studies show that governments could be getting higher revenues in the most famous and visited PAs, which could be then used to shore them up and/or supplement management needs elsewhere.

3. Discussion: future research needs

The Bruntland report [90] defined sustainable development as: "development that meets the needs of the present without compromising the ability of future generations to meet their needs." Phrased in more modern terms, it can be considered: "economic development that is conducted without depleting natural resources" (from Dictionary.com). In either case, many have questioned whether sustainable development can be achieved without stabilizing human populations and some consider that we are grossly overpopulated now, and beyond hope without massive population reductions (e.g. [91, 92]). We cannot address such global problems with our look at one sector of society in one country. Suffice it is to say that, like most developing countries of Asia and Latin America (but not Africa) Nepal's birthrate has declined greatly over the past three decades but it is still somewhat above the replacement rate. But here we can only focus on the topic at hand by asking how well Nepal has done in its conservation efforts, and where to go from henceforth.

We contend that, by most measures, Nepal has done very well in adopting modern conservation programs in a comparatively rather short time period. Core areas within the PA system now cover over 18% of the country's land area and, with buffer zones, the figure increases to over 20% (**Table 1**). With the implementation of community forestry, forest cover has increased across much of the country although biomass reduction is typical in nationalized forests outside PAs (e.g. [70, 93, 94]). More research is needed to determine how this affects plant and animal community composition and long-term forest health. On the managerial front, more coordination is needed between policies that involve MAPs harvest and those that involve community forests, CAs and BZs, where some extraction of common NTFPs is permitted (e.g. fodder, pole wood, fuel wood, etc.). While some efforts are underway, we contend that more needs to be done, especially given the physiographic variation of Nepal, and thus the differences in ecosystem productivity and composition due to altitudinal change on north-south axes and annual rain and snowfall averages on east-west and north-south axes (e.g. [95]). Progress has been made with partnerships between the department of plant resources and various NGOs and INGOs, as well as the forest department and various INGOs, but coordination between these related sectors has not. Ranabhat et al. [96] also made an argument in favor of policy coherence in Nepal's forestry sector and between forestry and other sectors.

While the PA system is now quite large, there are many latent issues that have been raised. Obvious successes have been that vast expanses of the major Himalayan peaks and valleys have been protected, most of Nepal's terai reserves were expanded in area and upgraded to NP status, and strides have been made in recognizing international reserves with both India and China. While gaps exist in these efforts (e.g. [97]) all represent advances and many wildlife populations have demonstrably increased despite growing human populations. But ecosystems and habitats in the middle hills of the country are under-represented in the PA system (e.g. [98–100]) and this remains a concern. The middle hills traditionally were the most populated rural parts of Nepal due to the difficulty of farming at high elevations or living year-round in the terai before malaria eradication. For these reasons, many species endemic to the middle hills are under greater threats than elsewhere in Nepal (e.g. [101]). Much more research is needed on wild populations in the middle hills outside PAs, and surveys should be undertaken to locate potential PAs in that region. Due to the abundance of private landholdings, any new PAs in the middle hills would necessarily be small. However, given what we now know about the diversity of small mammals, butterflies and plants in those areas, viable populations of many species of concern could likely be conserved within relatively small reserves (e.g. [2, 5, 15]).

With regard to climate change, in terms of research, mitigation and/or adaptation, Nepal presents huge potential concerns as well as opportunities, given its physiographic variability. There is great concern, for example, about impacts on Himalayan glaciers that have already begun and upon which millions of people throughout large portions of South Asia depend for potable water [18] to potential impacts on forests and freshwater wetlands due to changes in local and regional precipitation patterns [102]. Studies on assisted colonization are underway in many places and, given the vast biodiversity that Nepal harbors, there is no limit to other studies that could be done in-country. But this remains controversial (e.g. [17]) and some suggest that we should let nature take its course given that humans tend to focus only on species that are economically important or esthetically pleasing. Perhaps a more fruitful approach would be to move only species into well-protected habitat from which they were previously extirpated, as opposed to where they never occurred, which is what the DNPWC has done with rhino and buffalo, and attempted with swamp deer.

For the myriad of other species that may or may not thrive in places outside their known geographic ranges as climates change, perhaps the most efficacious rout would be to develop more potential north-south corridors for movement between PAs at different elevations. While this has been explored in the past, little progress has been made. Once again, this would require detailed surveys throughout forested areas of the middle hills to determine where core PAs could be established, albeit small in area as discussed above, and where and how they could be connected by corridors where feasible, or act as stepping stone reserves to enhance natural dispersal across elevations. A good deal of cross-sectorial coordination would be needed to eventuate these possibilities, but it would also need to begin with more-focused research.

There also remains a great deal of research to be done on policy and economics of conservation in Nepal. Studies on the ways in which community-level user groups work (or do not) within community forestry, BZ and CA organizations, for example, have led to some generalizations that may improve implementation over time (e.g. [68, 69]). But more studies are needed given the high socio-cultural diversity inherent to Nepal (e.g. [103]) and differences in economies and levels of development in different part of the country. There is vast potential for more ecological and environmental economics research in Nepal as well; studies are now underway that look explicitly at the values of forests, wetlands and some important species in the national economy and more efforts should be encouraged along these lines. The importance of the PA system to the tourism sector cannot be understated and this is another area with great research potential. The need for more infrastructure as tourist markets increase has been apparent in some places (e.g. [104]) as has the need for more and better-trained manpower and programs designed to train them [105]. Tourism brings money, money generates jobs, and more or better employment opportunities cause people to move, thereby increasing population densities around highly-visited PAs (e.g. [106]). Long-term sustainability is in question if more tourism leads to environmental degradation. Much more socio-economic research is needed on these topics, as well as on the issue of willingness-to-pay for entry fees that may lead to, for example, charging lower fees to encourage entries to little-visited PAs and higher fees to discourage excess entries elsewhere (e.g. [107, 108]).

There is no question that a global-scale human-induced mass extinction is underway and has been for quite some time (e.g. [109, 110]). There is also no question that the rate of extinction has increased greatly as human populations expanded from about 2 to about 7.5 billion over the past 100 years. In that time, Nepal's population is thought to have increased from several million to 30 million now, yet records are made regularly for species thought previously not to occur in the country (e.g. [111]) and the press regularly publishes articles on species newly discovered in Nepal and the Himalayan region (e.g. The Times of India, 11 February, 2017; The Daily Telegraph, 23 December, 2018). This is encouraging, as is the extent of the PA system, its rapid growth and the fact that much international conservation law is under implementation, enabling legislation has been drafted, and regulations have resulted from them that encourage sustainable resource use. The decrease in Nepal's population growth rate, the expansion of community-based conservation and the MAPs and wetland policies under implementation in recent decades, are all very encouraging. The amount of NGO and INGO involvement is also encouraging and has proven very effective for conservation. In addition, many wildlife populations are known to be increasing and forest cover has demonstrably increased across much of the country.

4. Conclusions

Does the formal conservation sector in Nepal meet standards of sustainable development via either of the definitions offered above? We contend that it does, with several caveats. For different countries, places and times, there are may be many roads to sustainable development. With regard to the conservation sector of Nepal, we argue that it is on one such road. The country has made far too many efforts and has had far too many successes to deny the obvious. But will and can it continue? The answer to that question lies outside the scope or purview of any one nation. But, we would also argue, with more research and the development and the implementation of more programs to solve difficult problems—such as likely consequences of global warming—Nepal has, at the very least, arguably set up an important and highly functional PA system, and a vastly-broader conservation sector, that is likely to be resilient in the face of change. Although much remains to be done, we are encouraged about the future of biodiversity conservation in Nepal in the rapidly-changing world of the Anthropocene.

Acknowledgements

We thank the officers and staff of the DNPWC and Departments of Forestry and Plant Resources who have facilitated our travels and permitted access to field sites as well as reports and other archived materials. We also thank colleagues and staff of the Institute of Forestry, the Central Departments of Botany and Zoology and the Natural History Museum of Tribhuvan University for many conversations and access to libraries and collections over the course of many years. Officers of staff of NGOs and INGOs have also graciously allowed us formal interviews, access to archived materials and freely gave of their time in myriad conversations about the conservation sector in Nepal. They include: WWF-Nepal, IUCN-Nepal, NTNC, The Mountain Institute, ICIMOD, The Annapurna Conservation Area Project, UNDP-Nepal, USAID-Nepal, the Nepal-Australia Forestry Project, Wildlife Conservation Nepal, Bird Conservation Nepal, Himalayan Nature, The Mountain Forum and the Nepal Forum for Environmental Journalists. Literally hundreds of people across many decades have helped us in our efforts. This review is dedicated to the memories of Dr. Pralad B. Yonzon, Dr. Tirtha Man Maskey, and Mr. Mingma N. Sherpa—among the greatest conservationists Nepal has ever known—who were taken far too soon.

Author details

Joel T. Heinen[1*], Nabin Baral[2], Prakash K. Paudel[3] and Jay P. Sah[4]

1 Department of Earth and Environment, Florida International University, Miami, FL, USA

2 University of Washington School of Environmental and Forest Sciences, Seattle, Washington, USA

3 Center for Conservation Biology, Kathmandu Institute for Applied Sciences, Kathmandu, Nepal

4 Southeast Environmental Research Center, Florida International University, Miami, FL, USA

*Address all correspondence to: heinenj@fiu.edu

References

[1] Heinen JT, Shukurov E, Sadykova C. Legislative and policy initiatives in biodiversity conservation in Kyrgyzstan. Post-Soviet Geography and Economics. 2001;**42**(7):519-543

[2] Chaudhary RP. Biodiversity in Nepal: Status and Conservation. Know Nepal Series; No. 17, S. Devi, Saharanpur, India. 1998. 324 p

[3] Heywood V, Davis SD. Centres of Plant Diversity: A Guide and Strategy for their Conservation. Vol. 2. Gland, Switzerland: IUCN/WWF; 1995

[4] Dimmick WW, Edds DR. Evolutionary genetics of the endemic schizorathicine fishes of Lake Rara, Nepal. Biochemical Systematics and Ecology. 2002;**30**(10):919-929

[5] Jnwali SR. The Status of Nepal's Mammals: The National Red List Series. Kathmandu, Nepal: IUCN—The World Conservation Union; 2007. 267 p

[6] Katuwal HM, Khanal B, Basnet K, Rai B, Devkota S, Rai SK, et al. The mammalian fauna from the central Himalaya, Nepal. Asian Journal of Conservation Biology. 2013;**2**(1):21-29

[7] Paudel PK, Bhattarai BP, Kindlmann P. An overview of the biodiversity in Nepal. In: Kindlmann, editor. Himalayan Biodiversity in the Changing World. The Netherlands: Springer; 2012. pp. 1-40

[8] Wikramanayake E, Dinerstein E, Loucks CJ, Morrison J, Lamoreux J, Mcknight M, et al. Terrestrial Ecoregions of the Indo-Pacific: A Conservation Assessment. Washington, DC: Island Press; 2002. 824 p

[9] Olson DM, Dinerstein E. The global 200: A representative approach to conservation the earth's most biologically valuable ecoregions. Conservation Biology. 1998;**12**:502-515

[10] Paudel PK, Heinen JT. Conservation at the crossroads: On the status of threatened fauna in the Central Himalayas of Nepal. Geoforum. 2015a;**64**:192-195

[11] Adhikari S, Baral H. Governing forest system services for sustainable environmental governance: A review. Environments. 2018;**5**(5):53. DOI: 10.3390/environments5050053

[12] Allendorf TD. Resident attitudes towards three protected areas in Southwest Nepal. Biodiversity and Conservation. 2007;**16**(7):2087-2102

[13] Mehta JN, Kellert SR. Local attitudes towards community-based conservation policy and programmes in Nepal: A case study from Makalu-Barun conservation area. Environmental Conservation. 1998;**25**(4):320-333

[14] Thing SJ, Paudel BS. Buffer zone community forestry in Nepal: Examining tenure options. Journal of Forest and Livelihood. 2017;**15**(1):57-70

[15] Anonymous. Proceedings of the First One-Day National Seminar on Small Mammals Issues. LDTA Building, Jawalakhel, Kathmandu, Nepal: Small Mammals Conservation and Research Foundation; 2010. 85 p

[16] Aryal A, Shrestha TK, Ram A, Frey W, Groves C, Hemmer H, et al. A call to save the wild water buffalo (*Bubalus arnee*) in Nepal. International Journal of Conservation Science. 2011;**2**(4):261-268

[17] Hunter ML. Climate change and moving species: Furthering the debate on assisted colonization. Conservation Biology. 2007;**21**(5):1356-1358

[18] Shrestha AB, Aryal R. Climate change and its impacts on Himalayan glaciers. Regional Environmental Change. 2011;**11**(1):65-77

[19] Heinen JT, Shrestha SK. Evolving policies for conservation: An historical profile of the protected area system of Nepal. Journal of Environmental Planning and Management. 2006;**49**(1):41-58

[20] Stevens SF. Tourism, change and continuity in the Mount Everest region, Nepal. Geographical Review. 1993;**83**(4):410-427

[21] Mishra HR. Balancing human needs and conservation in Nepal's Royal Chitwan National Park. Ambio. 1982;**11**(5):246-251

[22] Sharma UR. An overview of park-people interaction in Royal Chitwan National Park, Nepal. Landscape and Urban Planning. 1990;**19**(12):133-144

[23] Heinen JT. The faunal collapse of large mammals in the reserves of the Nepalese terai, Bangkok: FAO. The Tiger Paper. 1995a;**22**(4):18-24

[24] Dinerstein E. The Return of the Unicornis: The Natural History and Conservation of Greater One-Horned Rhinoceros in Nepal. New York, USA: Columbia University Press; 2003

[25] Bhattarai BP, Paudel PK, Kindlman P. Conservation of biodiversity: An outline of challenges. In: Kindlmann, editor. Himalayan Biodiversity in the Changing World. The Netherlands: Springer; 2012. pp. 41-70

[26] Bhattarai BR, Wright W, Paudel BS, Aryal A, Yadav B, Wagle R. Shifting paradigms for Nepal's protected areas: History, challenges and relationships. Journal of Mountain Science. 2017;**14**(5):964-979

[27] Budhathoki P. Developing Conservation Governance Strategies: Holistic Management of Protected Areas in Nepal [doctoral dissertation]. UK: University of Greenwich; 2012. 319 p. Available from: http://gala.gre.ac.uk/id/eprint/9453

[28] Heinen JT. Chapter 1: International trends in protected areas policy and management. In: Global Issues and Trends in the Protection of Natural Areas. Rijeka, Croatia: InTech; 2012. DOI: 10.5772/50061. 18 p

[29] Seidensticker J. On the ecological separation between tigers and leopards. Biotropica. 1976;**8**(4):225-234

[30] Smith JLD. The role of dispersal in structuring the Chitwan tiger population. Behavior. 1993;**124**(3):165-195

[31] den Braber B, Evans KL, Oldekop JA. Impact of protected areas on poverty, extreme poverty and inequality in Nepal. Conservation Letters. 2018;**11**(6):1-9. DOI: 10.1111/conl.12576

[32] Dahal S, Nepal S. Conservation for whom? Political Ecology and Tourism. 2016;**15**:130-144. DOI: 10.9774/gleaf.9781315723471_10

[33] Dewu S, Roskaft E. Community attitudes towards protected areas: Insights from Ghana. Oryx. 2018;**52**(3):489-496

[34] Heinen JT, Roque A, Collado-Vides L. Managerial implications of perceptions, knowledge, attitudes and awareness of Puerto Morelos reef National Park, Mexico. Journal of Coastal Research. 2017;**33**(2):295-303. DOI: 10.2112/coastres-d-15-00191.1

[35] Hochleithner S. Beyond contesting limits: Land, access and resistance at Virunga National Park. Journal of Forest and Livelihood. 2017;**15**(1):57-70

[36] Allendorf TD, Gurung B. Balancing conservation and development in

Nepal's protected area buffer zones. Parks. 2016;**22**(2):69-82

[37] Fox J, Yonzon PB, Podger N. Mapping conflict between biodiversity and human needs in Langtang National Park, Nepal. Conservation Biology. 1996;**10**(2):562-569

[38] Ma B, Zhoa Z, Ding H, Wen Y. Community attitudes towards protected areas: Case study of Sichuan giant panda reserves in China. Environment, Development and Sustainability. 2018;**20**(4):1665-1686

[39] Studsrod JE, Wegge P. Park-people relationships: The case of damage caused by park animals around the Royal Bardia National Park, Nepal. Environmental Conservation. 1995;**22**(2):133-142

[40] Sah JP, Heinen JT. Resource utilization and conservation attitudes among indigenous and migrant peoples in Ghodaghodi Lake, Nepal. Environmental Conservation. 2001;**28**(4):345-356

[41] Shrivastava RJ, Heinen JT. A microsite analysis of communities around Kaziranga National Park, India: Implications for conservation and development. The Journal of Environment and Development. 2007;**16**:207-216

[42] Acharya KP, Paudel PK, Neupane PR, Kohl M. Human-wildlife conflict in Nepal: Patterns of human fatalities and injuries cause by large mammals. PLoS One. 2016;**11**(9):e0161717. DOI: 10.1371/journal.pone.0161717

[43] Acharya KP, Paudel PK, Jnawali SR, Neupane PR. Can forest fragmentation and configuration work as indicators of human-wildlife conflict? Evidence from human deaths and injury by wildlife attacks in Nepal. Ecological Indicators. 2017;**80**:74-83

[44] Gurung B, Smith JLD, McDougal C, Karki JB. Factors associated with human-killing tigers in Chitwan National Park, Nepal. Biological Conservation. 2008;**141**(12):3069-3078

[45] Bhatt N. Kings as wardens and wardens as kings: Post-Rana ties between Nepali royalty and national park staff. Conservation and Society. 2003;**1**(2):247-268

[46] Lehmkuhl JF, Upreti RK, Sharma UR. National parks and local development: Grasses and people in Royal Chitwan National Park, Nepal. Environmental Conservation. 1988;**15**(2):143-148

[47] Sah JP. Koshi Tappu Wetlands: Nepal's Ramsar Site. Bangkok Thailand: IUCN; 1997. 254 p

[48] Nepal SK. Tourism and protected areas: The Nepalese Himalaya. Annals of Tourism Research. 2000;**27**(3):661-681

[49] Nepal SK, Chipeniuk R. Mountain tourism: Toward a conceptual framework. Tourism Geographies. 2005;**7**(3):313-333

[50] Nyaupane GP, Lew AA, Tatsugawa K. Perceptions of trekking tourism and social and environmental change in Nepal's Himalayas. Tourism Geographies. 2014;**16**(3):415-437

[51] Yonzon PB, Hunter ML. Cheese, tourists and red pandas in the Nepal Himalayas. Conservation Biology. 1991;**5**(2):196-202

[52] Bushley BR and Khatri DB. REDD+: Reversing, Reinforcing or Reconfiguring Decentralized Forest Governance in Nepal. Forest Action Nepal. Discussion Paper 11:3; Kathmandu, Nepal. 2011. 30 p

[53] Dongol Y, Heinen JT. Pitfalls of CITES implementation in Nepal: A policy gap analysis. Environmental Management. 2012;**50**(2):181-192

[54] Shrestha-Acharya R, Heinen JT. Emerging policy issues on non-timber forest products in Nepal. Himalaya. 2006;**26**(1-2):51-54

[55] Caplins C, Halvorson SJ, Bosak K. Beyond resistance: A political ecology of cordyceps as alpine niche product in Garhwal, India Himalaya. Geoforum. 2018;**96**:298-308

[56] Singh R, Kumar N, Yadav PK. Perspectives on medicinal plants and their role in the life of a rural community in Uttarakand, India. Journal of Biological Sciences and Medicine. 2018;**4**(1):16-26

[57] Baral HS and Inskipp C. Important Bird Areas in Nepal: Key Sites for Conservation. Bird Conservation Nepal. Report to the Royal Society for the Protection of Birds, UK. 2005

[58] Paudel N, Adhikari S, Paudel G. Ramsar lakes in the foothills of the Himalaya, Pokhara-Leknath, Nepal: An overview. Janapriya Journal of Interdisciplinary Studies. 2017;**6**:134-147

[59] Belbase N. The Implementation of Environmental Law in Nepal. Kathmandu, Nepal: IUCN; 1997. 125 p

[60] Heinen JT. International conservation agreements. In: Nierenberg WA, editor. Encyclopedia of Environmental Biology. Vol. 1. San Diego: Academic Press; 1995b. pp. 375-384

[61] Merriman JC, Gurung H, Adhikari S. Rapid ecosystem service assessment of Koshi Tappu wildlife reserve, Nepal, on wetland benefits to local communities. Wetlands Ecology and Management. 2018;**26**(4):491-507

[62] Baral N, Stern MJ, Bhattarai R. Contingent valuation in Annapurna conservation area, Nepal: Implications for sustainable park finance and local development. Ecological Economics. 2008;**66**(2):218-227

[63] Baral N, Stern MJ, Hammett AL. Developing a scale for evaluating ecotourism by visitors: A study in Annapurna conservation area, Nepal. Journal of Sustainable Tourism. 2012;**20**(7):975-989

[64] Das M, Chaterjee B. Ecotourism: A panacea or predicament? Tourism Management Perspectives. 2015;**14**:3-16

[65] Timilsina N, Heinen JT. Forest structure under different management regimes in the western lowlands of Nepal: A comparative analysis. Journal of Sustainable Forestry. 2008;**26**(2):112-131

[66] Baral S, Gautam AP, Vacik H. Ecological and economical sustainability assessment of community forest management in Nepal: A reality check. Journal of Sustainable Forestry. 2018;**37**(8):820-841

[67] Allendorf TD, Allendorf K. The role of gender in park-people relations in Nepal. Human Ecology. 2012;**40**(5):789-796

[68] Baral N, Gautam R, Timilsina N, Bhat MG. Conservation implications of contingent valuation of critically endangered white-rumped vultures (Gyps bengalensis) in South Asia. International Journal of Biodiversity Science and Management. 2007a;**3**(3):145-156

[69] Baral N, Stern M, Heinen JT. Integrated conservation and development project life cycles in the Annapurna conservation area, Nepal. Biodiversity and Conservation. 2007b;**16**(10):2903-2917

[70] Timilsina N, Ross MS, Heinen JT. A community analysis of Sal (Shorea robusta) forests in the western terai of Nepal. Forest Ecology and Management. 2007;**241**(1):223-234

[71] Belbase N. National Implementation of the Convention on Biological

Diversity. 1999. Available from: Agris.fao.rog/agris-search/searchdo?recordID=US201300061768

[72] Liu H, Gale SW, Cheuk ML, Fisher GA. Conservation impacts of commercial cultivation of endangered and overharvested plants. Conservation Biology. 2018:1-12. DOI: 10.1111/cobi.13216

[73] Lam JTL, Koldeway HJ, Yasue M, Vincent ACJ. Comparing interview and trade data in assessing changes in the seahorse, hippocampus spp., trade following CITES listing. Oryx. 2016;**50**(1):36-46

[74] Heinen JT, Chapagain D. The expansion of species protection in Nepal: Advances and pitfalls in new efforts to implement and comply with CITES. Journal of International Wildlife Law and Policy. 2002;**5**:235-250

[75] Heinen JT, Yonzon PB, Leisure B. Fighting the illegal fur trade in Kathmandu, Nepal. Conservation Biology. 1995;**9**(2):245-247

[76] Felbab-Brown V. Wildlife and drug trafficking, terrorism, and human society: Realities, myths and complexities beyond Africa. PRISM. 2018;**97**(4):124-137

[77] Hanson T. Biodiversity conservation and armed conflict: A warfare ecology perspective. Annals of the New York Academy of Sciences. 2018;**1429**(1): 50-65. DOI: 10.1111/nyas.13689

[78] Baral N, Heinen JT. The maoist people's war and conservation in Nepal. Politics and the Life Sciences. 2005;**24**(1-2):1-11

[79] Heinen JT, Rayamajhi S. On the use of goal-oriented project planning in Nepalese protected area management. Environmental Practice. 2001;**3**(4):227-236

[80] Heinen JT, Paudel P. On the translocation of Asian buffalo *Bubalus arnee* in Nepal: Are feral backcrosses worth conserving? Conservation Science. 2015;**3**(1):11-19

[81] Ale SB, Yonzon PB, Thapa K. Recovery of snow leopard in Sagarmatha (Mt. Everest) National Park, Nepal. Oryx. 2007;**41**(1):89-92

[82] Morrison JC, Sechrest W, Dinerstein E, Wilcove DS, Lamoreux JF. Persistence of large mammal faunas as indicators of global human impacts. Journal of Mammalogy. 2007;**88**:1363-1380

[83] Laiolo P. Diversity and structure of the bird community overwintering in the Himalayan subalpine zone: Is conservation compatible with tourism? Biological Conservation. 2004;**115**(2):251-262

[84] Aryal A, Brunton D, Raubenheiner D. Habitat assessment for the translocation of blue sheep to maintain a viable snow leopard population in the Mt. Everest region, Nepal. Zoology and Ecology. 2013;**23**(1):66-82

[85] Hanemann WM. Valuing the environment through contingent valuation. Journal of Economic Perspectives. 1994;**8**(4):19-43

[86] Lamsal P, Atreya K, Pant KP, Kumar L. Tourism and wetland conservation: Application of travel cost and willingness to pay an entry fee at Ghodaghai Lake complex, Nepal. Natural Resources Forum. 2016;**40**(1-2):51-61

[87] Dhital N. Reducing emission from deforestation and degradation (REDD) in Nepal: Exploring possibilities. Forests and Livelihood. 2009;**8**(1):57-62

[88] Baral N, Kaul S, Heinen JT, Ale S. Estimating the value of world heritage site designation: A case study from Sagarmatha (Mt Everest) National Park,

Nepal. Journal of Sustainable Tourism. 2017;**2017**(4):1-16

[89] Tobarra-Gonzalez MA, Mendoza-Monpean J. Recreational value of El Valle and Carascoy Natural Park. Anatolia. 2018. DOI: 10.1080/13032917.2018.1519196

[90] Anonymous. Our Common Future: The World Commission on Environment and Development. UK: Oxford University Press; 1987. 383 p

[91] Foreman D. Man Swarm: How Overpopulation Is Killing the Wild World. London, UK: Live True Books; 2015. 196 p

[92] Shragg S. Move Upstream: A Call to Solve Overpopulation. London, UK: Freethought House; 2015. 104 p

[93] Brant JS, Allendorf T, Radeloff V, Brooks J. Effects of national forest management regimes on unprotected forests in the Himalaya. Conservation Biology. 2017;**31**(6):1271-1282

[94] Khatri DBK. Comparison of forest biomass across a human-induced disturbance gradient in Nepal's Schima-Castinopsis forests. Journal of Sustainable Forestry. 1999;**9**(3-4):69-82

[95] Maren IE, Bhattarai KR, Chaudhary RP. Forest ecosystem services and biodiversity in contrasting Himalayan forest management systems. Environmental Conservation. 2014;**41**(1):73-83

[96] Ranabhat S, Ghate R, Bhatta LB, Agrawal NK. Policy coherence and the interplay between climate change adaptation policy and the forestry sector in Nepal. Environmental Management. 2018;**61**(6):968-980

[97] Kandel P, Gurung J, Chetri N, Ning W, Sharma E. Biodiversity research trends and gap analysis from a transboundary landscape, eastern Himalaya. Journal of Asia-Pacific Biodiversity. 2016;**9**(1):1-10

[98] Hunter ML, Yonzon PB. Altitudinal distribution of birds, mammals, people, forests and parks in Nepal. Conservation Biology. 1993;**7**(2):420-423

[99] Paudel PK, Heinen JT. Conservation planning in the Nepal Himalayas: Effectively (re) designing reserves for heterogeneous landscapes. Applied Geography. 2015b;**56**(1):127-134

[100] Shrestha UB, Shrestha S, Chaudhary P, Chaudhary RP. How representative is the protected area system of Nepal? A gap analysis based on geophysical and biological features. Mountain Research and Development. 2010;**30**(3):282-294

[101] Paudel PK, Sipos J, Brodie JF. Threatened species richness along a Himalayan elevational gradient: Quantifying the influences of human population density, range size and geometric constraints. BMC Ecology Series. 2018;**18**:6. DOI: 10.1186/s12898-018-0162-3

[102] Lamsal P, Kumar L, Atreya K, Pant KP. Vulnerability and impacts of climate change on forest and freshwater wetland ecosystems in Nepal: A review. Ambio. 2017;**46**(8):915-930

[103] Bista DB. Peoples of Nepal. 4th ed. Kathmandu: Nepal; Ratna Pustak Bhandar. 1981. 252 p

[104] Heinen JT, Thapa BB. A feasibility study of a proposed trekking trail in Chitwan National Park. Kathmandu: Tribhuvan University. Journal of Institute of Forestry. 1988;**10**:19-28

[105] Heinen JT. The design and implementation of a training program for tour guides in Royal Chitwan National Park, Nepal. Bangkok: FAO. The Tiger Paper. 1990;**17**(2):11-15

[106] Wittemyer G, Elsen P, Beau WT, Coleman A, Burton O, Brashares JS. Accelerated human population growth at protected area edges. Science. 2008;**32**:123-126

[107] Cuervorst RL, Lambron C. Visitor attitudes and value orientation for a proposed national monument. Journal of Outdoor Recreation and Tourism. 2018;**23**:33-43

[108] Job H, Becken S, Lane B. Protected areas in a neoliberal world and the role of tourism in supporting conservation and sustainable development: An assessment of strategic planning. Journal of Sustainable Tourism. 2017;**25**:1697-1718

[109] Kolbert E. The Sixth Extinction: And Unnatural History. New York: Holt and Company; 2014. 319 p

[110] Wilson EO. Half Earth: Our Planet's Fight for Life. New York, New York: Liveright Publishing Co; 2016. 274 p

[111] Heinen JT. Rare and new bird records for Kosi Barrage and Kosi Tappu wildlife reserve during winter and spring, 1987, Kathmandu: Tribhuvan Univ. Journal of Natural History Museum. 1988;**10**:23-30

Chapter 6

Parks and Reserves in Madagascar: Managing Biodiversity for a Sustainable Future

Patrick O. Waeber, Serge Rafanoharana, H. Andry Rasamuel and Lucienne Wilmé

Abstract

Madagascar has an extended network of over 100 protected areas with various IUCN status covering more than 10% of terrestrial landscapes and seascapes. The location of these areas is to a high-degree congruent with remaining forests covering some 15% of the island. The definitions of forests are numerous, at global, national, and regional scales; here we emphasize the widespread system considering the percentage of tree cover canopy, to better define the eastern humid vs. western dry forests in Madagascar and to understand how best to protect the terrestrial biodiversity within parks and reserves. Forests are home to over 80% of Madagascar's biodiversity. These ecosystems are under high threat due to ongoing and rapid deforestation and degradation. We present the interlinkages and complexity of governing National Parks to safeguard Madagascar's unique biodiversity and ecosystem services.

Keywords: forests, endemism, forest governance, protected area management, deforestation

1. Introduction

Madagascar is a large island in the Southwest Indian Ocean located east of South Africa and Mozambique. After its independence from former colonial France (1890–1960), the country has experienced more than five political crises since the early 1990s. Madagascar fares among the poorest countries, where over 80% of its 26 million inhabitants, mainly living in rural areas, are depending on less than 2$ per day. The grand majority of rural people depend on agricultural production—mainly rice, and maize in the drier parts—mostly for subsistence and to lesser extent for local markets.

Madagascar is also widely renowned for its unique biodiversity. The fourth biggest island, separated from mainland Africa some 165 million years ago (Ma), assembles a high degree of endemism at species and higher taxonomic level. With few exceptions, its hundreds of species of amphibian and reptiles and 100% of lemurs are occurring in Madagascar only. There is also a high degree of endemism—over 82%—in the >12,000 native species of vascular plants [1]. Many of these species are also evolutionary ancient going back millions of years [2], and there still remain several hundreds of species to be discovered and scientifically described.

Goes one of these endemics extinct on the island, it translates into a loss at global level. This makes conservation so much more challenging and pressing. Madagascar has also been declared as one of the world's 200 global hot spots for biodiversity due to its endemism and threat status [3].

The rich and unique biodiversity has attracted a lot of research and conservation attention of global order, bringing to the scene relatively large donor institutions [4, 5], also translating into a quite large conservation network of protected areas. To date, some 8.5% of terrestrial landscapes are under formal protection, and some 6.2% of Madagascar territorial waters are protected seascapes.

In this book chapter, we aim to show the importance of Madagascar's protected area network, which consists of over 120 sites. In the first section, we provide an overview of the biological richness, more than 80% of which depends on forest ecosystems. Forests themselves come along in a range of different types: from humid forests in the east to dry forests in the west, to dry spiny forests in the southwest. In the next section, we give a detailed account of the protected area system, its IUCN categories and evolution through time to reflect global changes and expectations (cf. [6]), and its governance. The third section will focus on challenges and threats to protected areas and their biodiversity and governance based on three selected case studies. The final section constitutes an outlook, where we reflect upon ideas of the most imminent actions needed for safeguarding the PA system and its unique biodiversity in a poverty and corruption-stricken country.

2. Forests and biodiversity

The biodiversity of Madagascar is renown and characterized by several features: (i) a high rate of endemism in species and at higher taxonomical levels including genera, families, and suborders and/or orders; (ii) some radiations in endemic families or subfamilies; (iii) monospecific or paucispecific endemic families or subfamilies, and (iv) the absence of certain groups well represented in Africa or Asia.

Research in paleontology carried out in Madagascar revealed Cretaceous birds, amphibians, snakes, and mammals. The peculiar *Beelzebufo* from the Upper Cretaceous (100.5–66 Ma) of western Madagascar was a giant frog more closely related to extant frogs present in South America but absent from Madagascar [7]. The Mesozoic (251.902–66 Ma) snakes recovered in Madagascar belong to an extinct family [8], while the modern mammals currently occurring in Madagascar have no ancestors among the known paleontological records. In the current state of our knowledge, with a few exceptions, the vertebrates currently occurring on the island have ancestors who arrived in Madagascar after the K-Pg mass extinction 66 million years ago (Ma) [9, 10].

The records of mammals show some recently extinct taxa, especially large-bodied lemurs. These include the 160 kg Sloth lemur *Archaeoindris fontoynontii*, more than 20 times heavier than the current larger lemurs, the 6–7 kg *Indri indri* or *Propithecus diadema*, or >5000 times the size of the smallest primate on Earth, the Madame Berthe's mouse lemur *Microcebus berthae*. The lemurs exhibit all types of activity patterns with most of the smaller species being mainly nocturnal, the largest species in the Indridae family being diurnal, and many species, mostly in the Lemuridae family being cathemeral with activities occurring throughout the 24 hours [11–13]. Few species of lemurs have a diet with a high proportion of fruits; their diet is always complemented by leaves, which seems to be related with low levels of nitrogen in the Malagasy fruits [14]. All lemur species but one occur in the various forest types of the island. The exception is the narrow-ranged Alaotra gentle lemur *Hapalemur alaotrensis* endemic to the wetlands of Lake Alaotra [15].

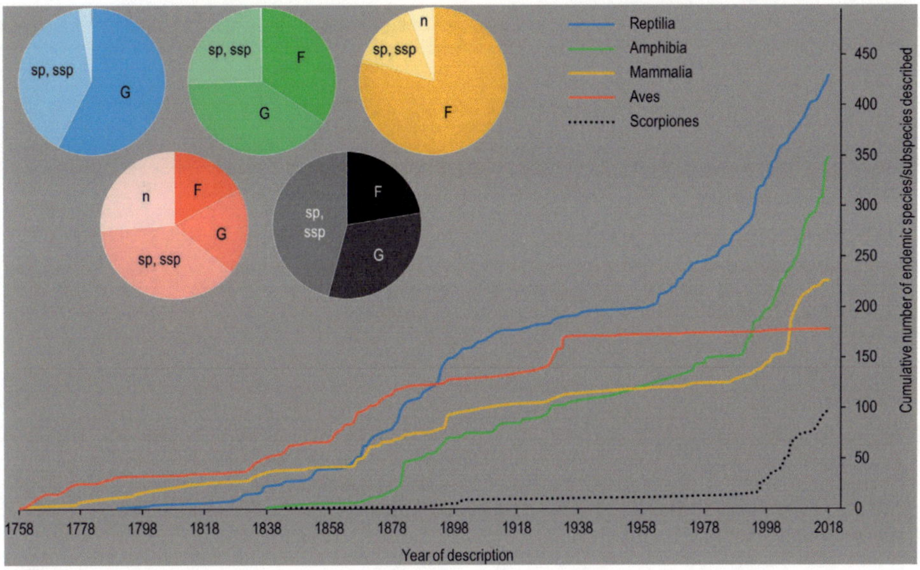

Figure 1.
Number of endemic species and subspecies described since 1758 and composition of the groups of scorpions, amphibians, reptiles, birds, and mammals in terms of level of endemism of the autochthonous species and subspecies found in Madagascar (F for species and subspecies belonging to an endemic family or subfamily, G for those belonging to an endemic genus; n for other native species and subspecies but not endemic).

The small terrestrial mammals of Madagascar are tenrecs and rodents. The tenrecs occur in all types of forest but have greater species richness in the humid forest. One species is solely encountered in rivers, the aquatic Web-footed tenrec *Limnogale mergulus*.

The humid forests encountered in the east and northwest harbor the highest richness of vertebrates, as compared to the western dry forests and the southern dry-spiny forest thickets. The endemic birds represent an exception; they are well represented in the driest forests of the west and the south [16]. The mammals, reptiles, and amphibians are represented by 191, 254, and 260 endemic species and subspecies, respectively; the birds have only 178 endemic species and subspecies (**Figure 1**, **Table 1**). These low numbers for birds are not fully understood, but recent research on the rich endemic scorpion fauna points toward a so-called "Neogrosphus rule" stipulating that "In a changing environment causing geographical barriers, the lower the species' dispersal ability, and the greater the niche breadth of the ancestor taxum, the higher the species richness" [17, 18].

The order of birds in Madagascar is thus both rich and poor: rich in endemic taxa with 178 endemic species and subspecies, but also one endemic order—or suborder according to authorities—three endemic families, one endemic subfamily, and 39 endemic genera. Several of these higher endemic taxa are rich and exhibit endemic radiations as the family Bernieridae or the subfamily Couinae with, respectively, 16 and 14 endemic species and subspecies [19–21]. The Malagasy avifauna can be considered as poor, because entire groups occurring in Africa are not represented in Madagascar or in extremely small numbers only. Small numbers can point toward recent colonization, as in the case of the Madagascar cisticola (*Cisticola cherina*), but not always. A great counterexample is the cuckoo-roller belonging to a monospecific family occurring in Madagascar and the Comoros only. There are very few endemic bird species linked to open grasslands: only one species of lark (*Eremopterix hova*) or a single species of cisticola, while about 50 species are known from continental Africa. Some 21 endemic species and subspecies (12.0%) are encountered in open

	Scorpiones	Amphibia	Reptilia	Aves	Mammalia
Order	—	—	—	1	—
Family	2	—	1	3	7
Subfamily	—	3	—	1	10
Genus	9	22	43	39	43
Subgenus	1	15	5	—	—
Species or subspecies	98	348	429	178	226
Nonendemic autochthonous species and subspecies	—	*1*	*12*	*63*	*13*
% of endemism of species and subspecies	100.0	99.7	97.3	73.9	94.6
Number of species or subspecies endemic at family or subfamily levels	22	120	1	41	189
Number of species or subspecies endemic at genus level	53	260	254	87	191
% of species and subspecies endemic at family or subfamily level	22.4	34.4	0.2	17.0	79.1
% of species and subspecies endemic at genus level	54.1	74.5	57.6	36.1	79.9

Table 1.
Number of endemic taxa in the order Scorpiones and four classes of vertebrates, namely amphibians, reptiles, birds, and mammals (nonendemic species and subspecies also indicated in italic for the purpose of calculations, numbers as of December 31, 2018, based on Noe4D [19]).

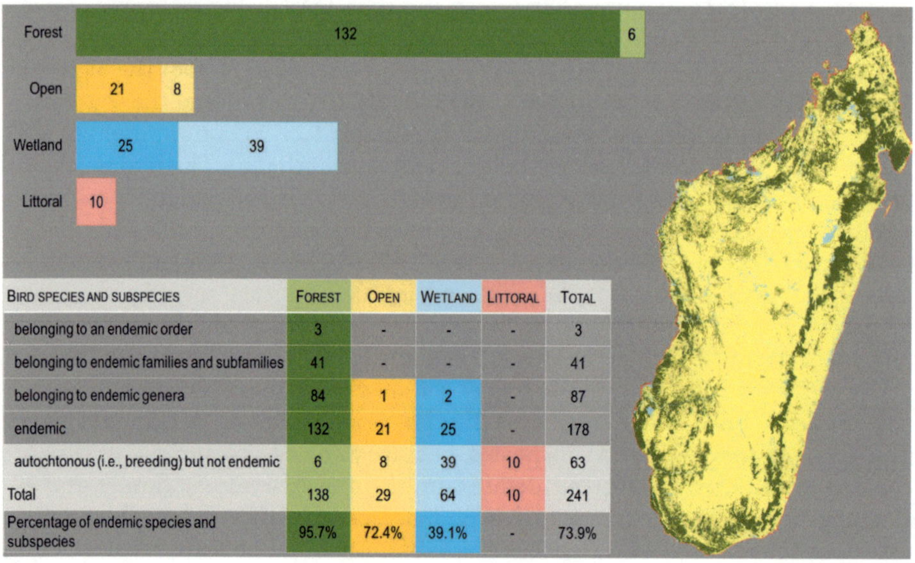

Figure 2.
Distribution of the endemic and nonendemic species and subspecies of birds according to the main type of habitat in which they occur (Madagascar simplified land cover based on [22]).

grasslands including the Madagascar partridge in the monospecific endemic genus *Margaroperdix madagarensis*, which is the only open grassland bird in a higher endemic taxon. Wetland birds are represented by 39 species and subspecies including 25 endemics (**Figure 2**). The paucity of endemic species in grasslands, some species encountered in wetlands, and the majority of species occurring in the different types of forests [22] fits the global pattern for the endemic vertebrates and invertebrates observed in Madagascar.

The largest bird in the World has been "made in Madagascar" and was much larger than the New Zealand Moas [23]. New Zealand and Madagascar, now >11,500 km apart, have some common geological history. Both islands were part of the Gondwana—separation from mainland Africa occurred at approximately 165 Ma, from Antarctica (and, indirectly, South America and Australia, linked with New Zealand) at approximately 115–112 Ma, and from the Indian subcontinent at approximately 88 Ma [10, 24]. These two large islands also share more recent avian history facts with the ratites. They were both home to giant birds that have disappeared in the last three to six centuries: the elephant birds in Madagascar and the Moas in New Zealand. It is widely reported that the elephant bird could have survived until the seventeenth century given that Etienne de Flacourt refers to it in his 1658 book [25]. Flacourt is also cited to refer to giant lemurs (e.g., [26]), on p. 154 of his book, thus encouraging some authors to give credit to the recent existence of these animals during the last centuries. The iconic New Zealand kiwis are also linked to Madagascar because they share a common ancestor with the elephant birds [27, 28]. Studies in New Zealand have revealed that humans were responsible for the extinction of the Moas some six centuries ago [29], but such firm conclusions have not been reached for the elephant birds or for the lemurs—contradicting views are still debated (e.g., [30, 31]). *Æpyornis titan* was described in 1894 [32] and later considered as a synonym of *Æ. maximus*. Hansford and Turvey [33] resurrected the species on the basis of new material recently collected in a cave of Isalo and transferred the species to a new genus Vorombe (Malagasy Vorona = bird and "be" = big). *Vorombe titan* is the largest bird that ever existed in the world. These giants weighing up to 860 kg were therefore seven to eight times heavier than the current ostriches of Africa. Recent studies have found that some elephant birds were vegetarian and nocturnal, a common trait shared with the New Zealand kiwis [34].

The elephant birds have also revealed the earliest known human presence in Madagascar [35]. In the oldest archeological site—the trading port Mahilaka in the northwest, established in the 10th century [36, 37]—human presence has more recently been depicted on animals with butchery marks on bones. For example, dwarf hippos, extinct giant lemurs, or elephants birds dated 2000 years before present BP [38, 39], or on a pigmy hippo bone dated 4000 BP [40]. Chicken bones are often used to date archeological sites (e.g., [41]); the chickens encountered in Madagascar are related to populations present in Indonesia and Africa [42]. A fossil Guinea fowl bone has been dated 10,000 years BP [43]; though Guinea fowl has better flight abilities than a chicken, it still cannot cross the >500 km distance separating Madagascar from Africa.

Box 1. *Elephant birds.*

The Central African forests are structured in such a way that a few dominant species and the large trees can explain the forests [44]. The tropical forests of western and central Africa are home to some 15,000 species of vascular plants (with an endemism estimated at 30%) of which 3000 are trees [45]. The countries from western and central Africa with tropical forests cover an area 12 times the size of Madagascar, which has nevertheless a comparative incredible richness of vascular plants, that is, more than 12,000 native plant species (82% are endemic to Madagascar), including more than 3000 native species of trees [1]. When compared to central Africa, Madagascar is peculiar due to the diversity of the types of vegetation encountered on the island.

Contrary to the Central African forests and other tropical forests in the World, the Malagasy forests are not renowned for their timber species with a few exceptions: the valuable timber species including the rosewood (*Dalbergia* spp.) and the ebony (*Diospyros* spp.) infamous for the traffic they are subject to [46, 47]. The high diversity encountered in the Malagasy flora is best explained by the diversity of vegetation types. Forests are ranging from humid forests in the east and the north where mean annual rainfall can reach 4000 mm to the southwestern spiny forest thickets in an extremely variable environment where mean annual rainfall can be below 500 mm [22, 48]. The central Menabe dry forests are structured with a few dominant species: the baobabs. The dry forests of western and southern Madagascar harbor six endemic species of baobabs (only one species occurs in Africa and one in Australia) [49].

Whether forests once covered all or almost the entire surface of the island remains controversial (e.g., [50]); however, it is certain that Madagascar has experienced major changes including during the recent Quaternary with its paleoclimatic oscillations. These oscillations and their effects on the abiotic factors, especially the numerous rivers with their watersheds, have allowed to propose

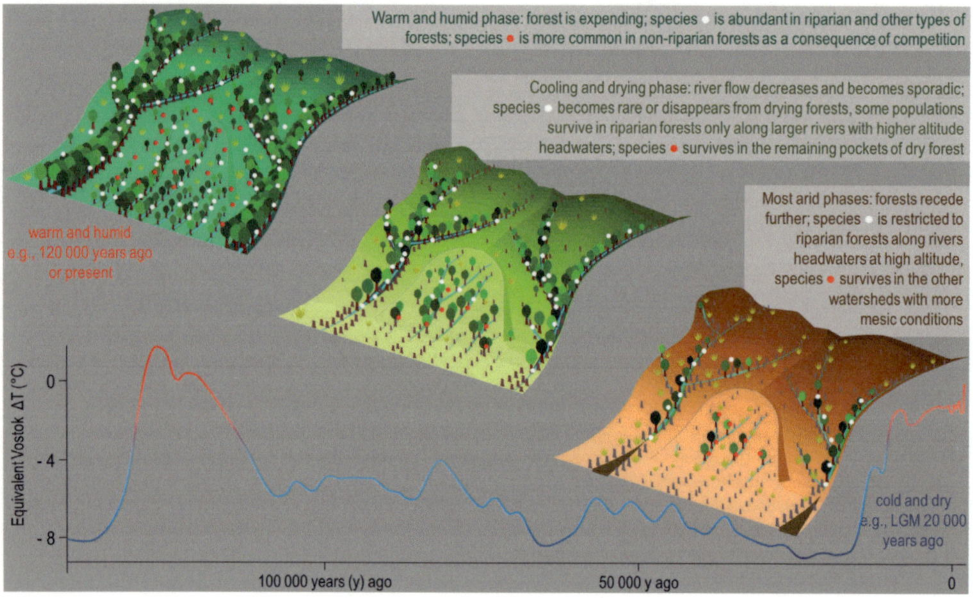

Figure 3.
Evolution of the abiotic factors, including the rivers and the slope, and biotic features, including the riparian forest and two faunal forest species, during paleoclimate oscillations (figure modified from [53], according to the Creative commons attribution 3.0 Unported license).

a model explaining the current distribution of some elements of the endemic forest biodiversity, more particularly the current range of some bird species, but especially those of many lemur species [51–54]. During the dry phases of the paleoclimate oscillations, the range of many populations were reduced to refugia, that is, places where water would still be available, therefore allowing the survival of a forest (vs. under high latitudes refugia are typically places free of ice). Finding out where these refugia could have occurred is part of the key to understand where the plants and animals were best in coping with the dry periods with decreasing rainfall (**Figure 3**).

3. Protected area system

Madagascar's high species richness and uniqueness has led to the creation of the first parks and reserves at the beginning of the twentieth century around the same time when the first protected areas were created on continental Africa [55]. The French administration established 10 Strict Nature Reserves at the end of 1927. By 1997, 46 protected areas (PA) covering almost 1.8 million hectares—with a clear focus on biodiversity conservation and research—were designated, including Strict Nature Reserve (IUCN Category I), National Parks (IUCN Category II), and Special Reserves (IUCN Category IV) [56–59] (**Figure 4**).

In the 1990s and 2000s, an important phase of internationally driven policy changes has been implemented in Madagascar. A suite of international conventions were signed and ratified laying ground for national legislative adaptation to better safeguard the biological and cultural patrimony and embed it in a global context [60]. The Convention on Biological Diversity CBD was ratified in 1995 (Law 1995-013 and Decree 1995-695); Ramsar Convention on Wetlands (Law 1998-003 and Decree 1998-261) was ratified in 1998; the Convention on the Conservation of Migratory Species of Wild Animals CMS was ratified in 2007. Madagascar was the

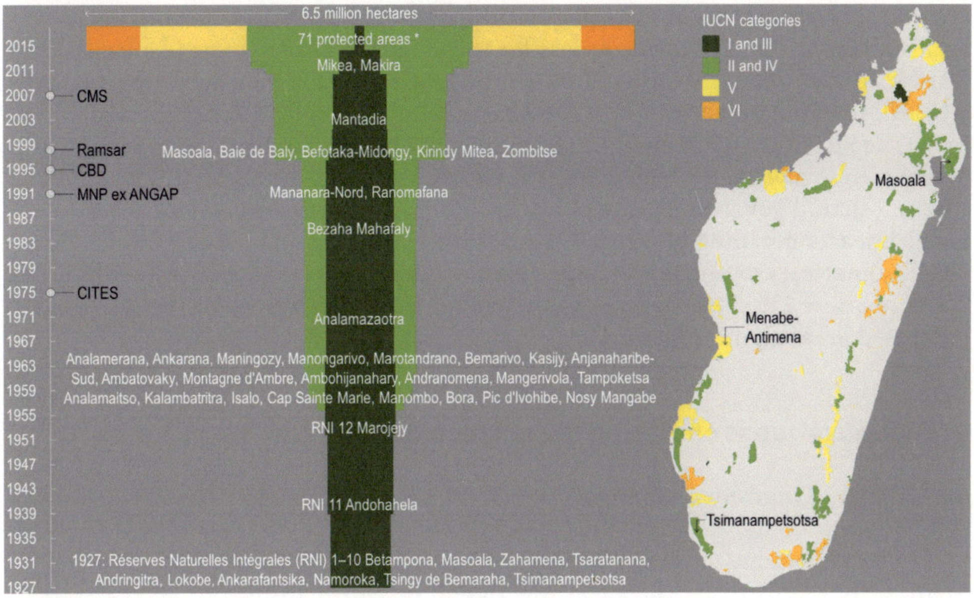

Figure 4.
*Protected areas and IUCN categories. Evolution of terrestrial protected areas in Madagascar (left), and distribution of the protected areas in 2018 (right). (*Agnakatrika, Agnalazaha, Alandraza Analavelo, Allée des Baobabs, Ambararata Londa, Ambatoatsinanana, Ambatofotsy, Ambatotsirongorongo, Ambodivahibe, Ambohidray, Ambondrombe, Ambositra-Vondrozo, Amoron'i Onilahy, Ampanangandehibe-Behasina, Ampasindava, Ampotaka-Ankorabe, Analabe Betanatanana, Analalava, Andrafiamena Andavakoera, Andreba, Angavo, Anjozorobe-Angavo, Ankarabolava, Ankarea, Ankeniheny-Zahamena, Ankivonjy, Ankodida, Antrema, Beanka, Behara Tranomaro, Bemanevika, Bombetoka Beloboka, Bongolava, COMATSA Nord, COMATSA Sud, Galoko-Kalobinono, Ibity, Itremo, Lac Alaotra, Loky Manambato, Mahavavy Kinkony, Mahialambo, Mahimborondro, Makirovana Tsihomanaomby, Mandena, Mandrozo, Mangabe-Ranomena-Sahasarotra, Mangoky Ihotry, Manjakatompo Ankaratra, Marolambo, Maromizaha, Menabe Antimena, Montagne des Français, Nord-Ifotaka, Nosy Antsoha, Oronjia, Petriky, Pointe à Larrée, Ranobe bay, Ranobe PK32, Rivière Nosivolo, Sahafina, Soariake, Sud-Ouest Ifotaky, Torotorofotsy, Tsimembo Manambolomaty, Tsinjoriake, Tsitongambarika, Velondriake, Vohidava-Betsimalaho, Vohidefo; CMS = Convention on the Conservation of Migratory Species of Wild Animals, Ramsar = Ramsar Convention on Wetlands, CBD = Convention on Biological Diversity, MNP = Madagascar National Parks, CITES = Convention on International Trade in Endangered Species of Wild Fauna and Flora).*

first country to develop and implement a roadmap for conservation and development. The New Environmental Action Plan (NEAP) brought about a decentralization in governance of natural resources management [61]. The GELOSE policy (Gestion Locale Sécurisée, secure local management) regulates the transfer of property rights from the State to local communities (Law 97-107 and Decree 97-1200); it is applicable to forests, pastures, water, and wildlife. Specifically, the protected area legislation Code des Aires Protégées (N. 848-05/N. 2001/05) legalized in 2003, and revisions related to the law of hunting (2006-098) and the Convention on International Trade in Endangered Species of Wild Fauna and Flora CITES (ratified in 1975) were undertaken to control the exploitation of wild animals and regulating in situ conservation (Decree 2006-400) (**Figure 4**). After the crisis years 2009–2013, the PA legislation code has been revised in 2015 under the Refonte du Code des Aires Protégées (N. 2015-005) and supplemented with an updated environmental charter, Charte de l'Environnement Malagasy actualisée (N. 2015-003).

During the fifth World Parks Congress 2003 (Durban, South Africa), President Marc Ravalomanana declared to triple the terrestrial surface under protection in Madagascar up to 6 million hectares under the so-called Durban Vision [62]. The new System of Protected Areas of Madagascar (SAPM)—established in 2002—including both, the management of the original PAs and new protected areas

(NPAs), is responsible for the safeguarding of biodiversity and cultural heritage and sustainably managing resources for the people of Madagascar (Commission SAPM, 2006). Until the early 2000s, the State through the parastatal organization ANGAP (created in 1991, now MNP, Madagascar National Parks) governed parks. Since the SAPM, management has become more diverse and complex, reflecting global trends toward protected area for people and biodiversity [63]. The NPAs follow a shared governance model by the regional government and local communities, accompanied by a legally recognized promoter (usually an international NGO, sometimes national NGO, universities, or even mining companies) [64, 65]. There are now 116 sites covering an area of 6.5 million hectares for the terrestrial protected areas alone (**Figure 4**).

4. Deforestation is emptying the protected areas

4.1 Drivers of land cover change

Madagascar's land cover has undergone big changes over the past decades. Many of the main drivers of change affecting forests are man-made in Madagascar. Agriculture is the number one promoter of change causing deforestation. In the tropics, agriculture is mainly characterized by family-based productions, operating in a dynamic mosaic of open landscapes dominated by grassy areas, slash, and burn shifting cultivation and land clearings [66]. In Madagascar, over 80% of the population lives in rural areas with some 78% of the active population engaging in agricultural activities [67].

Energy demand is another big driver across the island. Access to electricity is still poor, where some 15% of the total population—cities (37%) vs. rural inhabitants (4%)—benefit from it [68]. A grand majority of the 26 million inhabitants are still highly dependent on wood biomass for domestic energy supply (charcoal, firewood). An energy diagnostics conducted by WWF in 2012 revealed that more than 90% of rural inhabitants are still using fuel wood as main energy supply [69]. It is unknown to what extent the charcoal production is affecting natural forests within protected areas. While firewood collection affects any forests, charcoal production can take place in pine or eucalyptus plantations—which usually are grown on former forested lands, which is the case in many parts of the highlands—or it has been produced within natural forests. For example, in the southwest, production takes place mainly in the natural dry and spiny forests [70].

In contrast to other tropical countries, extractive industries such as oil and mining are of minor importance in Madagascar. While there is a number of prospecting projects ongoing, mainly on the western escarpment of the country—here lie the main natural geologically interesting layers (cf. [71])—there are only two large-scale mining operations installed. Small-scale artisanal (illegal) mining is much wider spread and oftentimes coincides with protected areas and forests. However, its impacts are far below the ones of agricultural expansion and energy needs. In contrast to some central African countries, there is no big-scale industrial logging.

4.2 Case studies

In the following, we present three protected areas from dense humid forests in the east, the National Park of Masoala, the western dry forest of the reserve of Menabe Antimena, to the southwestern dry spiny forest thicket of the National Park of Tsimanampetsotsa (**Figures 4** and **5**). We briefly depict its conspicuous biogeographic and ecological elements and summarize the main threats affecting these three sites.

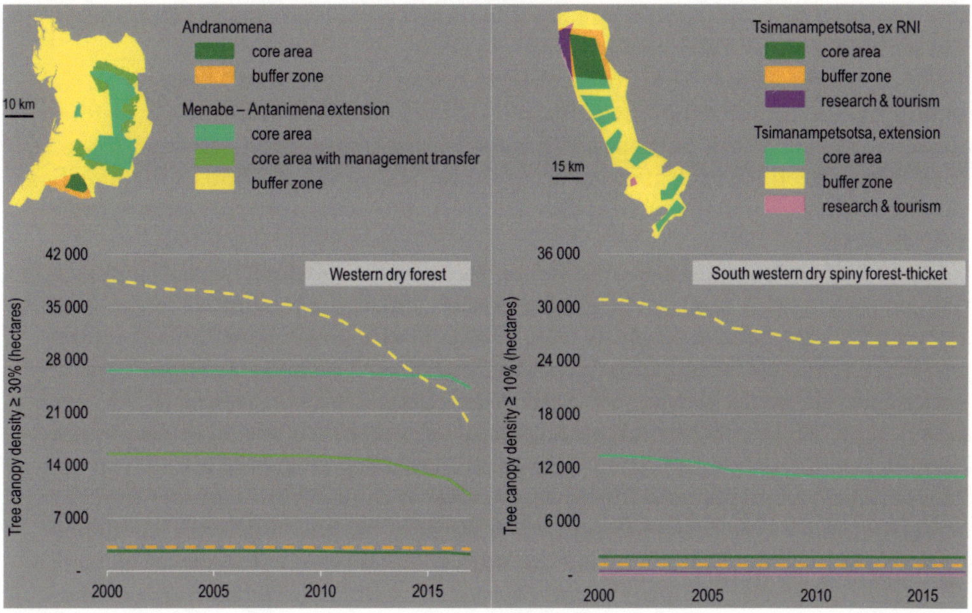

Figure 5.
Forest loss in Andranomena (left) and Tsimanampetsotsa (right) based on the different zoning of respective protected areas.

Case 1: The *Masoala National Park* with a total area of 212,000 hectares for its terrestrial portion alone is the largest National Park of Madagascar. It covers almost entirely the Masoala peninsula as well as the Center of endemism of Masoala [52]. The red ruffed lemur has its range limited to this center of endemism. The Strict Nature Reserve of Cap Masoala covering some 27,800 ha—created in 1927—has been degazetted in September 1964. The NP Masoala created in 1997 is unrelated to the previous RNI and protects the largest continuous humid forests of eastern Madagascar [72].

Main issues: The threats to the natural habitats and biodiversity of Masoala are deforestation and the permanent transformation of the habitat through slash and burn practices, the fragmentation of the habitat, and the overexploitation of some timber species, especially rosewood and ebony since the late 2000s [46, 72], and in places some overexploitation of tuber and animals [73].

Case 2: The *Menabe-Antimena reserve* with a total area of some 194,000 ha covers almost entirely the center of endemism of the Menabe between the large Mangoky River to the south and the Tsiribihina River to the north. The Menabe is a topography inclining toward the Mozambique Channel of hemicircular shape of about 200 km from north to south and 150 km from east to west. The Menabe is a kind of small sedimentary basin with three levels of cuestas that has emerged since the Miocene. The region experiences temperatures of 23–26°C, annual rainfall of 740 mm in Morondava, which increase inward and with altitude. The irregularity of the precipitations associated with a very strong evapotranspiration creates a hydrological deficit. All of the rivers within the Menabe are dry for several months a year between April and September. The most remarkable endemic species of the Menabe is probably the 2 kg giant jumping rat *Hypogeomys antimena* active at night, herbivorous and frugivorous, and monogamous [74, 75]. The forests of Madagascar are characterized by their biodiversity, including for the plants; if most forests cannot be seen through their large trees as in some parts of Africa, the Menabe forests are the exception with the baobabs (*Adansonia* spp.) dominating the canopy. The dry forests of Menabe-Antimena have a "natural" protection against fire: a natural fire would not enter deep into the forest layers.

Main issues: As for the majority of the western dry forests of Madagascar, the main threat encountered in the Menabe-Antimena is the destruction of the forest through slash and burn cultivation (**Figure 5**); especially in the past year, these forests have been burning as never seen before. Since the late 2000s, an intensification of the extraction of precious hardwood has been further adding pressures to the forest system [16].

Case 3: *The National Park of Tsimanampetsotsa* (also spelled Tsimanampesotse or Tsimanampetsotse) lies in the Center of endemism of the Karimbola in the driest region of Madagascar. As early as 1927, a Strict Nature Reserve of 17,520 hectares protected the Lake Tsimanampetsotsa and the spiny forest on its eastern bank toward the Mahafaly Plateau. The reserve was extended to an area of 43,200 hectares in June 1966 before encompassing a large proportion of the Mahafaly Plateau and the full diversity of its ecosystems in 2007, totaling an area of 203,400 hectares. The park lies in a region with the lowest annual rainfall of the country (<500 mm) with no organized river network, perched water table, and sink holes where blind gobies (*Typhleotris* spp.) inhabiting the subterraneous water system can be seen [76]. The park protects a narrow ranged endemic carnivore *Galidictis grandidieri*, two species of land tortoises, several species of lemurs and birds, and a dry spiny forest thicket with a high level of endemism.

Main issues: The Park of Tsimanampetsotsa is threatened by human activities, including overexploitation of natural resources such as increased cutting of natural forests (**Figure 5**) and trafficking of the radiated tortoises *Astrochelys radiata* [77, 78].

A common feature of threat from the two western sites is that the buffer zones are suffering most deforestation; also, the new core zones are more affected by anthropogenic activities. Interestingly, the old core zones and buffers experienced almost no deforestation. It seems that people are respecting the old boundaries. The newly added areas are more under pressure of agricultural needs from the adjacent riverine human populations as it seems less clear, which zones are exactly under what management regime. These confusions represent an important loophole in the governance system of the parks in Madagascar, which comes to the detriment of forests and biodiversity. Nevertheless, the forests in Madagascar would be very likely much more degraded and deforested without any formal and real protection; thus, these case studies can report at least some success in conservation and management efforts.

4.3 Methodology

For Madagascar's biodiversity analyses, we used the Noe4D database [19] comprising ca. 11,000 references and 52,800 georeferenced samples primarily documenting the endemic vertebrate fauna. We applied the layers provided in the Atlas of the vegetation of Madagascar [22] and the tree cover density Hansen/UMD/Google/USGS/NASA to evaluate forest cover and rates of deforestation on these three types of vegetation [79]. Data cleaning was conducted on the protected areas data sets, and the geometry was repaired due to self-intersections on each parcel of the data. Similar processing was done on the zonation of each protected areas, in addition to reprojection to similar coordinate system/projection. Vegetation data of Moat and Smith [22] and tree cover 2000 of Hansen [79] were clipped to the specified protected areas of interest, and both data sets were merged to get the type of vegetation per each pixel of the tree cover. The protected area zonation was then added to the merged data sets. Once the preparation of the data was completed, the tree loss data for the period 2001–2017 and the tree loss per each defined protected area zonation were calculated. Degraded humid forests on Masoala are significantly different from humid forests when a tree canopy density is higher than 90%. When we added the southwestern dry spiny forest-thicket layer on the tree cover density, we had to consider a tree cover density as low as 10% given that a tree cover density at 30% and higher did not reveal any forest cover; when choosing a tree cover density of 10% and above, the degraded southwestern dry spiny forest was also revealed (36.4% vs. 53.9%). The dry tropical forests are globally underestimated [80]. As shown in our case studies, the drier a forest, the higher the resulting underestimation of canopy cover. The southern and southwestern formations, because of their low stature and estimated low above ground carbon stocks, are of little interest for REDD+, which partly explains their low representativeness in the tree cover density's calculations. Nevertheless, these forests are extremely important in terms of endemic biodiversity, and we propose that a tree cover density of 10% be retained for this type of forest. Degraded dry forests cannot be distinguished from the intact dry forests [22]. Our analyses of the Menabe-Antimena suggest that a tree cover density not higher than 50% should be considered to evaluate the extent and quality of the Madagascar western dry forest.

5. Protected areas are gaining in importance

Deforestation in Madagascar has increased substantially in the past years [15]. Forests across Madagascar are being replaced for agricultural production, mainly through slash-burning, which does not halt at protected area boundaries. Fertility of freshly cleared forests yields production for few years only [81]. These lands are changing into grassland-dominated areas. This means that terrestrial biodiversity is increasingly restricted to protected areas, where some intact forests remain. The more the nonprotected areas are being denuded of their original vegetation cover, the more the protected areas are gaining importance for the survival of biodiversity.

The recent increase in the total area of protected areas in Madagascar (**Figure 4**) is clearly in line with the desire to protect Madagascar's biodiversity. The first stage of the process was completed in 2015 with the legalization of the status of these protected areas making the large expansion of the system. While Henri Humbert did not underestimate the complexity of protecting the natural wealth of Madagascar in 1946 [55], the current situation has certainly become more complex. To date, more threats are looming with a steadily and fast-growing human population in need of survival. Adding to the complexity is human-caused climate change, its impacts on biodiversity and ecosystem services are still to be researched in the case of Madagascar.

Several parks and reserves in Madagascar have been and are being researched with the aim to better understand how to protect the endemic biodiversity while resolving conflicts with the human populations relying on its ecosystem services for a livelihood. The parks and reserves selected in our case studies are the subject of research in this direction (e.g., SuLaMa in Tsimanampetsotsa, research by Harvard on Masoala, research by Durrell in Menabe).

The potential effects of anthropogenic climate change for Madagascar's biodiversity may likely be unparalleled in its recent geological history. As shown, the endemic biodiversity has experienced multiple paleoclimatic oscillations and other events that have shaped its biogeography. During the dry periods of recent paleoclimatic oscillations, there is no doubt that many taxa survived in refugia, typically places where vegetation and animals still had access to water. Populations of entire species have likely experienced some negative or positive fluctuations depending on their ability to cope with and adapt to dry conditions, as well as their ability to compete for scarce resources. The protected areas play an important role to ensure resilience of its biodiversity in the context of anthropogenic climate change. For this, a better understanding of how geomorphology, reliefs, watersheds, and paleo-refugia influenced the distribution of the endemic biodiversity is crucial.

5.1 Complexity bears challenges

The new protected areas phase has increased the structural complexity of governance, which came at the cost of efficiency and efficacy. Many authors have identified and listed factors that impact forest and protected area governance efficiency, such as financial and human resources constraints, or the remoteness and accessibility of protected areas or forests [4, 82–84]. Many of these aspects are of "technical" nature: (i) developing better management plans based on evidence rather than political or marketing reasons to attract further funding (e.g., [85]) that can lead to (ii) improved and more efficient implementation; (iii) capacity building of staff would, inter alia, also allow for (iv) a better optimization of the use of scarce financial and human resources. According to Mauvais, Coordinator of the IUCN Program on African protected areas, [86] "(...) improving governance will have an infinitely greater impact than just working on what we are doing or trying to do in the field," referring to the abovementioned, where technical hurdles are much easier to overcome, especially with the most recent developments in monitoring and in-time reporting of forest and park infractions (e.g., [87]). Tools like SMART (smart-conservationtools.org/) or novel ones like the GLAD alerts (https://glad.umd.edu/alerts), for example, allow the anticipation of forest fires, thus guiding management actions to address potential deforestation. The recording of forest soundscapes is also a promising and novel tool for monitoring biodiversity for conservation [88].

5.2 Implementing new policies

During the NEAP period in the 1990s, the international community spearheaded by the World Bank pushed the Malagasy government to implement forest governance devolution, that is, to better engage with riverine populations and have them engaged in the decision-making process and management of forests and protected areas. In rather a short time and with little evidence of success, more than 450 so-called transfers of management have been installed across Madagascar [89]. To date, it remains still unclear whether this comanagement has been fruitful [90, 91]. Main problems lie in the noncommunication between park agents and village representatives and the mutual nontrust in working together among other reasons (see [83] for an example of governance perceptions around the Zahamena National Park). As shown in our case study examples from the Tsimanampetsotsa or Menabe protected areas, core areas of the original parks (with higher restrictions) show much slower deforestation than areas under a comanagement agreement.

Before rushing into new policy implementations to mirror global trends, utmost caution needs to rule to best assess potential risks. The CBD's (Convention on Biological Diversity) Aichi-Targets (target 11) require that by 2020, "at least 17% terrestrial and inland water areas and 10% of coastal and marine areas, (...) are conserved." While the protected area approach is globally still the most powerful conservation tool to safeguard biodiversity (e.g., [92, 93]), a problem in Madagascar with increasing the protected area surfaces, both on land and sea, is that there is no automatic guarantee of increased protection. Rather, this will increase the already gargantuan task of governing and managing protected areas and its forests and biodiversity. Currently, there are over 1 million ha of protected areas (26 sites) of so-called "paper parks," that is, not managed at all [61]. One looming factor, regardless of governance and management, is the financing of these parks. If the international community is interested in safeguarding the unique biodiversity, which represents both, a national and global patrimony, then it should ensure the proper funding of it. Madagascar alone cannot stem the financial burden of the protected areas network [61].

6. Conclusions

The area of protected areas has been greatly increased over the past years. The protected areas represent last vestiges for intact forests, as fragmentation and degradation are advancing at fast pace mainly outside their borders. Governance has now the gargantuan task to ensure that the parks and reserves are fulfilling their role of protecting the endemic biodiversity of Madagascar. The endemic biodiversity is still far from being known, research needs to be maintained to document but also to adapt the network of protected areas to allow that the entire biodiversity can benefit from protection, as well as find adequate and necessary means to carry out this task (viz., conservation and management tools, and funds, to list but the most important once). The endemic biodiversity of Madagascar is an inestimable heritage for the generations of Madagascar and the world, and the parks and reserves are its best chance for the future.

Acknowledgements

This research is part of the GEF-funded project Global Forest Watch in Madagascar (ID: 5356).

Conflict of interest

We declare no conflict of interests.

Author details

Patrick O. Waeber[1*], Serge Rafanoharana[2], H. Andry Rasamuel[2] and Lucienne Wilmé[2*]

1 Forest Management and Development, Department of Environmental Sciences, Swiss Federal Institute of Technology Zurich, Zurich, Switzerland

2 World Resources Institute, Antananarivo, Madagascar

*Address all correspondence to: patrick.waeber@usys.ethz.ch and lucienne.wilme@wri.org

References

[1] Phillipson PB, Andriambololonera S, Lowry II PP, Manjato N, Rabarimanarivo M, Rakotonirina N, Ravololomanana N, Schatz GE. The Madagascar Catalogue, progress to date and prospects for the future. In: Abstracts, XXI AETFAT Congress; Nairobi, Kenya; 2017. p. 241

[2] Crottini A, Madsen O, Poux C, Strauß A, Vieites DR, Vences M. Vertebrate time-tree elucidates the biogeographic pattern of a major biotic change around the K-T boundary in Madagascar. Proceedings of the National Academy of Sciences of the United States of America. 2012;**109**:5358-5363. DOI: 10.1073/pnas.1112487109

[3] Ganzhorn JU, Lowry PP, Schatz GE, Sommer S. The biodiversity of Madagascar: One of the world's hottest hotspots on its way out. Oryx. 2001;**35**:346-348. DOI: 10.1046/j.1365-3008.2001.00201.x

[4] Waeber PO, Wilmé L, Mercier J-R, Camara C, Lowry PP II. How effective have thirty years of internationally driven conservation and development efforts been in Madagascar? PLoS One. 2016;**11**:e0161115. DOI: 10.1371/journal.pone.0161115

[5] Corson C. A history of conservation politics in Madagascar. Madagascar Conservation & Development. 2017;**12**:49-60. DOI: 10.4314/mcd.v12i1.4

[6] Jenkins CN, Joppa L. Expansion of the global terrestrial protected area system. Biological Conservation. 2009;**142**:2166-2174

[7] Evans SE, Groenke JR, Jones MEH, Turner AH, Krause DW. New material of *Beelzebufo*, a hyperossified frog (Amphibia: Anura) from the Late Cretaceous of Madagascar. PLoS One. 2014;**9**:e87236. DOI: 10.1371/journal.pone.0087236

[8] Pritchard AC, McCartney JA, Krause DW, Kley NJ. New snakes from the upper cretaceous (Maastrichtian) Maevarano formation, Mahajanga basin, Madagascar. Journal of Vertebrate Paleontology. 2014;**34**:1080-1093. DOI: 10.1080/02724634.2014.841706

[9] Longrich NR, Tokaryk T, Field DJ. Mass extinction of birds at the Cretaceous-Paleogene (K-Pg) boundary. Proceedings of the National Academy of Sciences. 2011;**108**:15253-15257. DOI: 10.1073/pnas.1110395108

[10] Samonds KE, Godfrey LR, Ali JR, Goodman SM, Vences M, Sutherland MR, et al. Imperfect isolation: Factors and filters shaping Madagascar's extant vertebrate fauna. PLoS One. 2013;**8**:e62086. DOI: 10.1371/journal.pone.0062086

[11] Kappeler PM, Ganzhorn JU. The evolution of primate communities and societies in Madagascar. Evolutionary Anthropology: Issues, News, and Reviews. 1993;**2**:159-171

[12] Wright PC. Lemur traits and Madagascar ecology: Coping with an island environment. American Journal of Physical Anthropology. 1999;**110**(S29):31-72

[13] Eppley TM, Donati G. Cathemeral. In: Vonk J, Shackelford TK, editors. Encyclopedia of Animal Cognition and Behavior. Cham AG, Switzerland: Springer; 2019. DOI: 10.1007/978-3-319-47829-6_246-1

[14] Donati G, Santini L, Eppley TM, Arrigo-Nelson SJ, Balestri M, et al. Low levels of fruit nitrogen as drivers for the evolution of Madagascar's primate communities. Scientific Reports. 2017;**7**:14406. DOI: 10.1038/s41598-017-13906-y

[15] Waeber PO, Schuurman D, Wilmé L. Madagascar's rosewood (*Dalbergia* spp.) stocks as a political challenge.

PeerJ Preprints. 2018:27062. DOI: 10.7287/peerj.preprints.27062v1

[16] Waeber PO, Wilmé L, Ramamonjisoa B, Garcia C, Rakotomalala D, Rabemananjara ZH, et al. Dry forests in Madagascar, neglected and under pressure. International Forestry Review. 2015;**17**(S2):127-148. DOI: 10.1505/146554815815834822

[17] Lourenço WR, Wilmé L, Waeber PO. More about the geographical distribution of the Malagasy genus *Neogrosphus* Lourenço, 1995 (Scorpiones: Buthidae) and description of a vicariant new species. Comptes Rendus Biologies. 2015;**38**:768-776. DOI: 10.1016/j.crvi.2015.08.001

[18] Lourenço WR, Wilmé L, Waeber PO. More about the geographical pattern of distribution of the genus *Pseudouroplectes* Lourenço, 1995 (Scorpiones: Buthidae) from Madagascar. Comptes Rendus Biologies. 2016;**339**:37-43. DOI: 10.1016j. crvi.2015.11.001

[19] Waeber PO, Gardner CJ, Lourenço WR, Wilmé L. On specimen killing in the era of conservation crisis—A quantitative case for modernizing taxonomy and biodiversity inventories. PLoS One. 2017;**12**:e0183903. DOI: 10.1371/journal. pone.0183903

[20] Goodman SM, Wilmé L. Cuculiformes: *Coua* spp. In: Goodman SM, Benstead JP, editors. The Natural History of Madagascar. Chicago: University of Chicago Press; 2003. pp. 1102-1108, 1147-1158

[21] Hasegawa M, Arai E. Rufous coloration associated with terrestrial locomotion in the adaptive radiation of Malagasy Couas. Journal of Ornithology. 2016;**157**:1115-1118. DOI: 10.1007/s10336-016-1370-2

[22] Moat J, Smith P, editors. Atlas of the Vegetation of Madagascar. Atlas de la Végétation de Madagascar. Kew: Kew Publishing, Royal Botanic Gardens; 2007. 124 p. Available from: http://www.vegmad.org/datasets.html

[23] Alexander RMN. Allometry of the leg bones of moas (Dinornites) and other birds. Journal of Zoology. 1983;**200**:215-231. DOI: 10.1111/j.1469-7998.1983.tb05785.x

[24] Ali JR, Krause DW. Late Cretaceous bioconnections between Indo-Madagascar and Antarctica: Refutation of the Gunnerus Ridge causeway hypothesis. Journal of Biogeography. 2011;**38**:1855-1872. DOI: 10.1111/j.1365-2699.2011.02546.x

[25] de Flacourt E. Histoire de la Grande Isle Madagascar. Troyes: Nicolas Oudot; 1658. p. 471 (i–xxi)

[26] Godfrey LR, Jungers WL. The extinct sloth lemurs of Madagascar. Evolutionary Anthropology. 2003;**12**:252-263. DOI: 10.1002/evan.10123

[27] Mitchell KJ, Llamas B, Soubrier J, Rawlence NJ, Worthy TH, Wood J, et al. Ancient DNA reveals elephant birds and kiwi are sister taxa and clarifies ratite bird evolution. Science. 2014;**344**: 898-900. DOI: 10.1126/science.1251981

[28] Yonezawa T, Segawa T, Mori H, Campos PF, Hongoh Y, Endo H, et al. Phylogenomics and morphology of extinct paleognaths reveal the origin and evolution of the ratites. Current Biology. 2017;**27**:68-77. DOI: 10.1016/j. cub.2016.10.029

[29] Gemmell NJ, Schwartz MK, Robertson BC. Moa were many. Proceedings of the Royal Society of London B: Biological Sciences. 2004;**271**:S430-S432. DOI: 10.1098/ rsbl.2004.0234

[30] Dewar RE, Richard AF. Madagascar: A history of arrivals, what happened, and will happen next. Annual Review of Anthropology. 2012;**41**:495-517. DOI: 10.1146/annurev-anthro-092611-145758

[31] Crowley BE, Godfrey LR, Bankoff RJ, Perry GH, Culleton BJ, Kennett DJ, et al. Island-wide aridity did not trigger recent megafaunal extinctions in Madagascar. Ecography. 2017;**40**: 901-912. DOI: 10.1111/ecog.02376

[32] Andrews CW. III—Note on a new species of *Æpyornis* (*Æ Titan*). Geological Magazine. 1894;**1**:18-20. DOI: 10.1017/S0016756800140373

[33] Hansford JP, Turvey ST. Unexpected diversity within the extinct elephant birds (Aves: Aepyornithidae) and a new identity for the world's largest bird. Royal Society Open Science. 2018;**5**:181295. DOI: 10.1098/rsos.181295

[34] Torres CR, Clarke JA. Nocturnal giants: Evolution of the sensory ecology in elephant birds and other palaeognaths inferred from digital brain reconstructions. Proceedings of the Royal Society B. 2018;**285**:20181540. DOI: 10.1098/rspb.2018.1540

[35] Lawler A. Scarred bird bones reveal early settlement on Madagascar. Science. 2018;**361**:1059. DOI: 10.1126/science.361.6407.1059

[36] Mahilaka RC. An archaeological investigation of an early town in northwestern Madagascar. Studies in African Archaeology. 1998;**15**:1-293. Available from: http://www.arkeologi.uu.se/digitalAssets/32/32403_3mahilika.pdf

[37] Crowther A, Lucas L, Helm R, Horton M, Shipton C, Wright HT, et al. Ancient crops provide first archaeological signature of the westward Austronesian expansion. Proceedings of the National Academy of Sciences. 2016;**113**:6635-6640. DOI: 10.1073/pnas.1522714113

[38] MacPhee RDE, Burney DA. Dating of modified femora of extinct dwarf *Hippopotamus* from southern Madagascar: Implications for constraining human colonization and vertebrate extinction events. Journal of Archaeological Science. 1991;**18**:695-706. DOI: 10.1016/0305-4403(91)90030-S

[39] Perez VR, Godfrey LR, Nowak-Kemp M, Burney DA, Ratsimbazafy J, Vasey N. Evidence of early butchery of giant lemurs in Madagascar. Journal of Human Evolution. 2005;**49**:722-742. DOI: 10.1016/j.jhevol.2005.08.004

[40] Gommery D, Ramanivosoa B, Faure M, Guérin C, Kerloc'h P, Sénégas F, et al. Les plus anciennes traces d'activités anthropiques de Madagascar sur des ossements d'hippopotames subfossiles d'Anjohibe (Province de Mahajanga). Comptes Rendus Palevol. 2011;**10**:271-278. DOI: 10.1016/j.crpv.2011.01.006

[41] Wilmé L, Waeber PO, Ganzhorn JU. Human translocation as an alternative hypothesis to explain the presence of giant tortoises on remote islands in the Southwestern Indian Ocean. Journal of Biogeography. 2017;**44**:1-7. DOI: 10.1111/jbi.12751

[42] Razafindraibe H, Mobegi VA, Ommeh SC, Rakotondravao J, Bjørnstad G, Hanotte O, et al. Mitochondrial DNA origin of indigenous Malagasy chicken: Implications for a functional polymorphism at the Mx gene. Annals of the New York Academy of Sciences. 2008;**1149**:77-79. DOI: 10.1196/annals.1428.047

[43] Goodman SM, Rakotozafy LMA. Subfossil birds from coastal sites in western and southwestern Madagascar. In: Goodman SM, Patterson BD, editors. Natural Change and Human Impact in Madagascar. Washington, DC: Smithsonian Institution Press; 1997. pp. 257-279

[44] Bastin J-F, Barbier N, Réjou-Méchain M, Fayolle A, Gourlet-Fleury S, Bogaert J, et al. Seeing Central African forests through their largest trees.

Scientific Reports. 2015;**5**:13156. DOI: 10.1038/srep13156

[45] Sosef MSM, Dauby G, Blach-Overgaard A, van der Burgt X, Catarino L, et al. Exploring the floristic diversity of tropical Africa. BMC Biology. 2017;**15**:15. DOI: 10.1186/s12915-017-0356-8

[46] Randriamalala H, Liu Z. Rosewood of Madagascar: Between democracy and conservation. Madagascar Conservation & Development. 2010;**5**:11-22. DOI: 10.4314/mcd.v5i1.57336

[47] Wilmé L, Waeber PO. Madagascar: Guard last of the forests. Nature. 2019;**565**:567. DOI: 10.1038/d41586-019-00323-6

[48] Dewar RE, Richard AF. Evolution in the hypervariable environment of Madagascar. Proceedings of the National Academy of Sciences of the United States of America. 2007;**104**:13723-13727. DOI: 10.1073/pnas.0704346104

[49] Petignat A, Jasper L. Baobabs of the World. The Upside-Down Trees of Madagascar, Africa and Australia. South Africa: Struik Nature; 2016. 112 p

[50] McConnell WJ, Kull CA. Protecting lemurs: Madagascar's forests. Science. 2014;**344**:358. DOI: 10.1126/science.344.6182.358-a

[51] Wilmé L, Goodman SM, Ganzhorn JU. Biogeographic evolution of Madagascar's microendemic biota. Science. 2006;**312**:1063-1065. DOI: 10.1126/science.1122806

[52] Wilmé L, Ravokatra M, Dolch R, Schuurman D, Mathieu E, Schuetz H, et al. Toponyms for centers of endemism in Madagascar. Madagascar Conservation & Development. 2012;**7**:30-40. DOI: 10.4314/mcd.v7i1.6

[53] Mercier J-L, Wilmé L. The eco-geo-Clim model: Explaining Madagascar's endemism. Madagascar Conservation & Development. 2013;**8**:63-68. DOI: 10.4314/mcd.v8i2.3

[54] Waeber P, Wilmé L, Mercier JL, Rakotozafy LMA, Garcia C, Sorg JP. The role of lakes in the context of the centers of endemism. Akon'ny Ala. 2015;**32**:34-47. Available from: https://essaforets.wordpress.com/akonny-ala-le-journal-forestier-malgache/n31-foret-et-recherche/

[55] Humbert H. La protection de la nature à Madagascar. Journal of the Arnold Arboretum. 1946;**27**:470-479

[56] Bruner AG, Gullison RE, Rice RE, da Fonseca GAB. Effectiveness of parks in protecting tropical biodiversity. Science. 2001;**291**:125-128

[57] Rasoavahiny L, Andrianarisata M, Razafimpahanana A, Ratsifandrihamanana AN. Conducting an ecological gap analysis for the new Madagascar protected area system. Parks. 2008;**17**:12-21

[58] Dudley N, editor. Guidelines for Applying Protected Area Management Categories. Switzerland: IUCN, Gland; 2008. x + 96 p. Available from: https://portals.iucn.org/library/sites/library/files/documents/PAG-021.pdf

[59] Gardner CJ, Nicoll ME, Mbohoahy T, Oleson KLL, Ratsifandrihamanana AN, Ratsirarson J, et al. Protected areas for conservation and poverty alleviation: Experiences from Madagascar. Journal of Applied Ecology. 2013;**50**:1289-1294. DOI: 10.1111/1365-2664.12164

[60] Rakotoarivelo AR, Razafimanahaka JH, Rabesihanaka S, Jones JPG, Jenkins RKB. Lois et règlements sur la faune sauvage à Madagascar: Progrès accomplis et besoins du futur. Madagascar Conservation & Development. 2011;**6**:37-44. DOI: 10.4314/mcd.v6i1.68063

[61] Mercier J-R. Madagascar moving towards sustainable development. The preparation of the National Environmental Action Plan (NEAP): Was it a false start. Madagascar Conservation & Development. 2006;**1**:50-54

[62] Norris S. Madagascar defiant. Bioscience. 2006;**56**:960-965. DOI: 10.1641/0006-3568(2006)56[960:MD]2.0.CO;2

[63] Mace GM. Ecology: Whose conservation? Science. 2014;**346**: 1558-1560. DOI: 10.1126/science.1254704

[64] Alvarado ST, Buisson E, Carrière SM, Rabarison H, Rajeriarison C, Andrianjafy M, et al. Achieving sustainable conservation in Madagascar: The case of the newly established Ibity Mountain protected area. Tropical Conservation Science. 2015;**8**:367-395. DOI: 10.1177/194008291500800207

[65] Gardner CJ, Nicoll ME, Birkinshaw C, Harris A, Lewis RE, Rakotomalala D, et al. The rapid expansion of Madagascar's protected area system. Biological Conservation. 2018a;**220**: 29-36. DOI: 10.1016/j.biocon.2018.02.011

[66] Bowman DMJS, Haberle SG. Paradise burnt: How colonizing humans transform landscapes with fire. Proceedings of the National Academy of Sciences of the United States of America. 2010;**107**:21234-21235. DOI: 10.1073/pnas.1016393108

[67] UNICEF. 2011. Available from: https://www.wfp.org/sites/default/files/MAG%202010%20CFSVA+N_Full%20report_English.pdf

[68] CIA World Factbook. 2018. Available from: https://www.cia.gov/library/publications/the-world-factbook/geos/ma.html [Accessed: January 10, 2019]

[69] WWF. 2013. Available from https://bit.ly/2E83dsx [Accessed: January 10, 2019]

[70] Dirac C, Andriambelo L, Sorg J-P. New research project in Central Menabe. Scientific bases for a participatory forest landscape management. Madagascar Conservation & Development. 2006;**1**: 31-33. DOI: 10.4314/mcd.v1i1.44041

[71] Raharimahefa T. Geoconservation and geodiversity for sustainable development in Madagascar. Madagascar Conservation & Development. 2012;**7**:126-134. DOI: 10.4314/mcd.v7i3.5

[72] Kremen C, Razafimahatratra V, Guillery RP, Rakotomalala J, Weiss A, Ratsisompatrarivo J-S. Designing the Masoala National Park in Madagascar based on biological and socioeconomic data. Conservation Biology. 1999;**13**:1055-1068. DOI: 10.1046/j.1523-1739.1999.98374.x

[73] Borgerson C, Fisher RN. Optimizing conservation policy: The importance of seasonal variation in hunting and meat consumption on the Masoala peninsula of Madagascar. Oryx. 2016;**50**:405-418. DOI: 10.1017/S0030605315000307

[74] Sommer S. Monogamy in *Hypogeomys antimena*, an endemic rodent of the dry deciduous forest in western Madagascar. Journal of Zoology. 1997;**241**:301-314. DOI: 10.1111/j.1469-7998.1997.tb01961.x

[75] Sommer S, Hommen U. Modelling the effects of life-history traits and changing ecological conditions on the population dynamics and persistence of the endangered Malagasy giant jumping rat (*Hypogeomys antimena*). Animal Conservation. 2000;**3**:333-343. DOI: 10.1111/j.1469-1795.2000.tb00118.x

[76] Vences M, Rasoloariniaina JR, Riemann JC. A preliminary assessment of genetic divergence and distribution of Malagasy cave fish in the genus *Typhleotris* (Teleostei: Milyeringidae). Zootaxa. 2018;**4378**:367-376. DOI: 10.11646/zootaxa.4378.3.5

[77] Manjoazy T, Razafimanahaka JH, Ronto W, Randrianavelona R, Ganzhorn JU, Jenkins RKB. The supply of illegal tortoise meat to Toliara City, South-Western Madagascar. Oryx. 2017;**51**:437-440. DOI: 10.1017/S0030605316000314

[78] Nopper J, Lauströer B, Rödel M-O, Ganzhorn JU. A structurally enriched agricultural landscape maintains high reptile diversity in sub-arid South-Western Madagascar. Journal of Applied Ecology. 2017;**54**:480-488. DOI: 10.1111/1365-2664.12752

[79] Hansen MC, Potapov PV, Moore R, Hancher M, Turubanova SA, et al. High-resolution global maps of 21st-century forest cover change. Science. 2013;**342**:850-853. DOI: 10.1126/science.1244693. Available from: https://earthenginepartners.appspot.com/science-2013-global-forest/download_v1.5.html [Accessed: 2019-01-10]

[80] Bastin J-F, Berrahmouni N, Grainger A, Maniatis D, Mollicone D, Moore R, et al. The extent of forest in dryland biomes. Science. 2017;**56**:635-638. DOI: 10.1126/science.aam6527

[81] Gay-des-Combes JM, Robroek BJM, Herve D, Guillaume T, Pistocchi C, Mills RTE, et al. Slash-and-burn agriculture and tropical cyclone activity in Madagascar: Implication for soil fertility dynamics and corn performance. Agriculture, Ecosystems & Environment. 2017;**239**:207-218. DOI: 10.1016/j.agee.2017.01.010

[82] McConnell WJ, Sweeney SP. Challenges of forest governance in Madagascar. Geographical Journal. 2005;**171**:223-238. DOI: 10.1111/j.1475-4959.2005.00162.x

[83] Bodonirina N, Reibelt LM, Stoudmann N, Chamagne J, Jones TG, Ravaka A, et al. Approaching local perceptions of forest governance and livelihood challenges with companion modeling from a case study around Zahamena National Park, Madagascar. Forests. 2018;**9**:624. DOI: 10.3390/f9100624

[84] Vieilledent G, Grinand C, Rakotomalala FA, Ranaivosoa R, Rakotoarijaona JR, Allnutt TF, et al. Combining global tree cover loss data with historical national forest-cover maps to look at six decades of deforestation and forest fragmentation in Madagascar. Biological Conservation. 2018;**222**:189-197. DOI: 10.1016/j.biocon.2018.04.008

[85] Gardner CJ, Waeber PO, Razafindratsima OH, Wilmé L. Decision complacency and conservation planning. Conservation Biology. 2018;**32**:1469-1472. DOI: 10.1111/cobi.13124

[86] Mauvais G. Available from: https://ideas4development.org/en/protected-areas-africa-governance/ [Accessed: January 10, 2019]

[87] Hansen MC, Krylov A, Tyukavina A, Potapov PV, Turubanova S, Zutta B, et al. Humid tropical forest disturbance alerts using Landsat data. Environmental Research Letters. 2016;**11**:034008. DOI: 10.1088/1748-9326/11/3/034008

[88] Burivalova Z, Game ET, Butler RA. The sound of a tropical forest. Science. 2019;**363**:28-29. DOI: 10.1126/science.aav1902

[89] Horning NR. Across the great divide: Collaborative forest management. In: Horning NR, editor. The Politics of Deforestation in Africa: Madagascar, Tanzania, and Uganda. 1st ed. Basingstoke, UK: Palgrave McMillan; 2018. pp. 135-163

[90] Rasolofoson RA, Ferraro PJ, Jenkins CN, Jones JP. Effectiveness of community forest management at reducing deforestation in Madagascar. Biological

Conservation. 2015;**184**:271-277. DOI: 10.1016/j.biocon.2015.01.027

[91] Rasolofoson RA, Ferraro PJ, Ruta G, Rasamoelina MS, Randriankolona PL, Larsen HO, et al. Impacts of community forest management on human economic well-being across Madagascar. Conservation Letters. 2017;**10**:346-353. DOI: 10.1111/conl.12272

[92] Polak T, Watson JEM, Bennett JR, Possingham HP, Fuller RA, Carwardine J. Balancing ecosystem and threatened species representation in protected areas and implications for nations achieving global conservation goals. Conservation Letters. 2016;**9**:438-445. DOI: 10.1111/conl.12268

[93] Kearney SG, Adams VM, Fuller RA, Possingham HP, Watson JEM. Estimating the benefit of well-managed protected areas for threatened species conservation. Oryx. 2018. DOI: 10.1017/S0030605317001739

Chapter 7

Noninvasive Sampling: Monitoring of Wild Carnivores and Their Parasites

Laís Verdan Dib, João Pedro Siqueira Palmer, Camila de Souza Carvalho Class de Lima, Otilio Machado Pereira Bastos, Claudia Maria Antunes Uchôa, Maria Regina Reis Amendoeira, Augusto Cezar Machado Pereira Bastos and Alynne da Silva Barbosa

Abstract

This chapter aims to present the importance, advantages, and disadvantages as well as the different types of noninvasive samples that can be used to monitor the carnivorous fauna and the parasitic agents that can infect these animals. This issue is extremely relevant, since noninvasive sampling has been increasingly used in different scientific researches that study animals with elusive habits, such as carnivores, and that claim animal welfare, once these animals do not need to be observed or captured. It is still important to highlight the scarcity of studies on parasitic diseases in free-living carnivores, being needed that parasitological surveys be done frequently by the conservation unit managers also to monitor the infectious agents that may be being introduced into the ecosystem of carnivores due to anthropization.

Keywords: gastrointestinal parasites, wild carnivores, coproparasitologic, trichology, molecular biology

1. Introduction

1.1 Animal identification from noninvasive samples

The study of free-living wild animals is a challenge for researchers for several reasons, including obtaining biological samples from these animals. There are three main types of sampling: destructive sampling, which is a strategy whose biological samples, mainly tissue, are obtained from animals that have been killed; nondestructive sampling in which the animal is normally captured and biopsy or

IntechOpen

blood collection techniques are performed in an invasive manner; and noninvasive sampling, that is, a strategy in which biological samples are obtained without the capture or manipulation of the animal [1].

Noninvasive samples are traces left by the animals in the places where they live, including hairs and loose feathers, feces, and other remnants of the diet [1, 2]. Noninvasive sampling is a strategy widely used by researchers in field studies, mainly biologists, since this method allows studies of free-ranging animals without the need to capture, manipulate, or even observe them [1, 3]. In this context, the analysis of noninvasive samples becomes an alternative with great cost-benefit for monitoring and, consequently, for the conservation of species, mainly free-living animals with nocturnal, elusive habits and that present low population densities, like carnivores, and those living in places of difficult access [4].

Among the different types of noninvasive samples, feces are ideal tools for indirectly analyzing free-ranging wild animals and the rocky areas, dry as well as frozen ecosystems provide the best conditions for stool preservation [4, 5]. By analyzing fecal material, it is possible to obtain information about the natural environment, including identification of the species that inhabits the region, composition of its diets, the function of that animal in the ecosystem, such as seed dispersal or population control of other animals, data on the taxa of prey ingested, especially in the case of carnivorous and omnivorous animals, and on the dynamics of gastrointestinal parasites in the environment and between animals; this is one of the major causes of mammalian fauna decline.

Thus, animal identification from the feces collected in the environment is very important and is possible by means of macroscopic analysis of fecal material [4], trichology of guard hairs [6], or using molecular biology techniques for the detection of animal DNA [1].

One of the ways of identifying animal taxa from noninvasive samples is by analyzing the morphology of feces or diet remains in these samples. The evidenced dietary components, such as claws, bones, teeth, and feathers, as well as the shape, size, and odor of feces are peculiar characteristics that can differentiate some animal groups. The use of morphological analysis in the study of feces of free-living wild animals is very advantageous, especially in field work, since it serves as an initial screening of the samples to be collected, allowing the classification of feces reliably, at least up to the category of order [7]. According to this author, this type of analysis is not such a safe resource for classification of samples up to family taxonomy, much less in gender or species. One of the disadvantages of using macroscopic analysis is that feces of free-living carnivores are exposed to suboptimal conditions for long periods of time and under the influence of different environmental circumstances, which may contribute to the loss of physical characteristics of the material and compromise a more reliable analysis of the sample [4].

In general, feces produced by carnivorous species have a cylindrical shape, are long (sausage type) with subdivisions, and presented one of the sharp ends. In the case of felids, in addition to the characteristics described above, it can be observed macroscopically that the feces tend to be more compact and have well-defined subdivisions and one of the ends is especially tapered and even slightly twisted. Stool diameter is also a very important feature to consider when estimating the size of the animal, that is, and to distinguish feces from small and large felids. In America, when fecal material is larger than 2.1 cm in diameter, they generally belong to large felids, such as *Puma concolor* and *Panthera onca* [4]. However, it is important to emphasize that the size and quantity of fecal cake produced by carnivores vary according to their age and type of feed intake. Therefore, large feces of carnivores that are still cubs can be easily confused with fecal material of carnivores of small size, for example.

Large carnivores are territorialistic animals, and when they are in reproductive or emotional times, they use different strategies to demarcate their territory [8, 9]. Usually, the territory marking is done through the urine (odoriferous marking), being this an important carrier of chemical information, or feces. A curious fact is that wild felines, such as *Panthera*, *Puma*, and *Leopardus*, do not bury their feces, contrasting with habits performed by domestic cats, which defecate in places where they can be buried [8]. This can be explained, since the burying of feces is related to the dominance or subordination of the animal. In this context, wild felines, which are dominant animals, do not hide their feces, leaving them visible for the demarcation of their territory [10].

Omnivorous animals' feces, such as canids, mustelids, and procyonids, fruit remains, seeds, insects, crustaceans, and plant remains are most commonly found. It is also worth mentioning that the fecal material of canids has a characteristic odor and, in the case of some species of canids such as *Chrysocyon brachyurus*, it has been verified that the diameter of the fecal material has been presented much larger than the feces of great felids, being these potential factors for the differentiation of these animal groups [4].

Using macroscopic analysis, the stool should be weighed with the analytical balance. Afterward, it is relevant that the researcher registers the color, presence of artifacts, and possible components of the diet, as well as the measurement of the length and diameter of all fecal propagules collected with the aid of a pachymeter. Subsequently, all the information obtained needs to be compared with the literature for the taxonomic classification of the fecal material author. Since there is little information about mammals' stool morphology and measurements in the literature, the comparison of all the feces collected is really difficult, especially in the cases of feces belonging to small neotropical wild felids.

Another mean of identifying animals from noninvasive sampling is analyzing lost hairs left in environment by trichology techniques. Hairs are keratinized epidermal attachments characteristic exclusively of mammals, being the second part of the body of the animal with greater durability [11]. Macroscopically, it is possible to distinguish two major regions in the hairs: the shield, characterized by being a longer and thicker distal region of the hair, and the stem, which is the sharpened portion and close to the bulb [12]. Morphologically, the hairs are composed of three layers: the cuticle, which is the outermost part; the cortex, which is the middle layer; and the medulla, which constitutes the innermost portion of the hairs [13, 14]. The cuticle consists of superimposed transparent keratin scales. From the base of the hairs, the distal portion of each scale lies on the proximal portion of the scale located above. Due to this conformation, the hairs have less resistance at their base when compared to their distal end [12]. The classification of the cuticular pattern can be established by the analysis of the imbrications, ornamentation, and continuity of the edges of the scales, shape, dimension, and oration of the scales [6]. In relation to the cortex, its thickness is what determines the width of the hair, being formed by keratinized, fusiform, small, and coalescent cells in a quasi-homogeneous hyaline mass with vacuoles and pigmented granules that can be organized as an amorphous mass [13]. Since the medulla is composed in a similar way to the cortex, however, its cells are clearly visible. Cells and air-filled spaces between intracellular connections are responsible for conferring the marrow characteristic [12]. The medulla can be classified according to its presence and continuity, number of rows, disposition and shape of the cells, and ornamentation of marrow margin [6].

The coat of most mammals is basically composed of two distinct types of hairs: the guard hairs, also called overhairs, which are the longest, smooth, and usually much pigmented, and the underhairs which are finer, shorter, and less pigmented and can be curved or curled [15]. In general, underhairs are more numerous and cover most of the animal body; therefore, they have as main functions the mechanical protection against impacts and the thermal regulation of the mammalian body. The guard hairs mainly present sensorial functions, being constituents of antennas, mustaches, vibrissae, and other regions with tactile functions of the mammals [14].

Some characteristics of guard hairs are used to identify mammalian species, since the combination of the cuticle, medulla, and cortex presents morphological patterns that confer specific diagnostic characteristics to a particular mammal species [6]. The microstructure of guard hairs is a useful tool in the identification of mammalian species, being applied several areas of research such as forensic science, ecology, epidemiology, archeology, and paleontology [6]. In this context, guard hair analysis is increasingly being used by researchers in mammalian ecology studies, mainly in the identification of the predator and diet analysis from noninvasive samples collected in the environment [11]. In the case of carnivores, especially the felids, self-cleaning as a habit of corporal hygiene allows that in their fecal material, a great quantity of hairs, including those by the guard, are evidenced. In addition, the study of hairs deposited in museum collections in order to help the understanding and standardization of the nomenclature of the cuticular and spinal patterns, as well as the use of this material as reference, has been increasingly adopted in studies with noninvasive free-living animal samples [11]. It should be emphasized that the guard hairs of the ear, head, neck, paws, and tail have different microstructural characteristics from those of the hairs of the rest of the body, which are the majority, being detrimental to identifications.

The cuticle scales vary in size and shape depending on the region of the hairs being analyzed. Normally, at the extremities of the scales, they are small in size, whereas in the wider portion of the shield, the scales are larger and are arranged transversely to the larger axis of the hair. In the stem, in turn, a greater variety of cuticular patterns occur, being this a region of high diagnostic value and the best part of the hair for the differentiation of groups or species of animals. As for the medulla, the best region of the hair for its observation would be the broadest part of the shield [12].

In order to perform trichology, it is necessary to separate a portion of each sample [6]. First, the fecal samples have to be submitted to the washing, drying, and storage stages in the laboratory. The washing step is done in a sink with running water and the aid of fine-mesh tampons with 1 μm diameter for the separation of any type of hair and removal of the fecal material remaining. All hairs can be placed on a sheet of white paper labeled with the sample number for drying at 37°C in an oven. After drying, all the hairs, including the guard hairs, are stored in satin plastic bags. Afterward, the guards were separated on a white surface, which can be another sheet of paper, using two tongs, on a white-lined bench. The hairs can be separated into individual plastic bags according to their length and morphological similarities. In addition, it is also important to separate artifacts such as claws, feathers, seeds, scales, small bones, and other nondigestible materials. Then, some selected guard hairs recovered from the fecal samples are submitted to cuticle printing and medulla diafanization in order to find the predator's guard hair [6].

Among the advantages offered by trichology, also observed by our research group, is the identification of the animal species, both of the predator and of possible prey [6, 16]. In addition, it is possible to obtain this material for analysis from noninvasive samples, mainly feces, with no manipulation or encounter of the animal needed to obtain hairs. Another factor that contributes to the performance of trichology in the field of research is its cost-benefit, since the reagents and utensils used can be obtained easily and at a low price, such as nail polish, lathe, plastic bags, slides, and commercial hydrogen peroxide [16]. The disadvantages faced are the difficulty of recovery of the predator's guards, since the volume of ingested hairs is significantly lower when compared to the ingestion of a prey, especially rodents. In this context, the macroscopic selection of the guardians of predators for analysis can often be ineffective. In the case of carnivores that eat other animals of the same order, trichology becomes a complementary tool, and other techniques are necessary for a more specific diagnosis. In addition, since the hairs are very delicate objects, it is necessary that the room where the researcher is working be isolated in order to prevent the hairs from dispersing in the environment [17]. Another obstacle would be the difficulty in producing high-quality slides for a reliable diagnosis of the species, especially with hairs obtained from fecal samples, that are often deteriorated or fragmented [6, 12, 15]. The most difficult cuticula pattern for our group to print was those with pavement wave type of scale imbrication, and the easiest ones were the hairs composed by imbricated foliaceous scales.

A third method that can be used in mammals' identification is the DNA analysis by molecular biology. The use of molecular markers for the study of free-living wild animals from noninvasive samples such as feces and dietary components has been increasingly applied in the research field, especially among carnivorous species with low population densities [1, 18]. From the molecular analysis, it is possible to obtain precise taxonomic information on the species, sexing, ecology, distribution, population estimates, and behavior of these animals, including their eating habits, reproductive preferences, and the pathogens that may be infecting these animals [1, 19, 20]. The main sources of DNA obtained from noninvasive samples are hairs, feces, urine, feathers, snake scales, skins, eggshells, and even skeletons. In the case of DNA analyses from fecal material, studies have shown that colon wall epithelial cells eliminated by the animal at the time of defecation are reliable sources of genetic material for identification and investigation of other information on the feces author [19, 21].

In relation to the molecular markers used in noninvasive samples for identification of the animal species, several primers have already been described, and many of them have been adapted, mainly from mitochondrial genes. Some characteristics of mitochondrial DNA such as the absence of recombination, high rate of evolution, and the large number of copies in the cell are the main advantages of its use as molecular marker, unlike nuclear genes [22]. The first molecular markers used to identify animal species were those named in the "universal primers" literature that amplify homologous fragments of several species, such as cytochrome b (CytB), which amplify fragments of 307 base pairs [21, 23, 24]. Another primitive also used was the cytochrome C and oxidase I subunit (COI), which amplifies about 650 base pairs and which was initially described to identify insects but which has also been widely used for the study of vertebrates [25]. In the case of animals inserted at high levels of the food chain, "universal primers" are poorly indicated for the identification of the predator because they also amplified nucleotide fragments of other animals, such as prey, and are therefore nonspecific [26–28]. All over the years, the mitochondrial genome has been extensively studied in the free-living mammals, such as the 16S region [28], the control region [29], ATP6 [28, 30], and 12S [27]. These genetic markers enhance the chances of success in polymerase chain reaction (PCR), since they amplify smaller DNA fragments and increases the probability of degraded DNA detection in noninvasive samples [28].

Despite all the advantages provided by the molecular methods for the study from noninvasive samples of wild animals in free life, these also present a series of limitations. Some of the obstacles faced in obtaining the DNA sequences of interest are the low quantity and low quality of genetic material in the samples, the extraction, and amplification method employed. This is because normally noninvasive samples of wild free-living animals are in the environment exposed to climate conditions, which may cause degradation of the genetic material present therein [1]. Since the genetic material is very deteriorated, it is important not to dilute with pure water the reagents and the carnivore DNA in the tube when performing the PCR. The use of water to complete the required volume in several standard protocols can lead to a lower sample amplification rate. In accordance with our experience, it is therefore recommended that the water be completely withdrawn so as to increase the chances of DNA application in noninvasive samples and to obtain a minimum volume required for PCR, purification, and sequencing. In addition, the presence of genetic material from other organisms, such as prey, mainly on samples of large carnivores, plants in the case of omnivorous or herbivorous animals, as well as bacteria and fungi, may produce nonspecific bands or even void the amplification of samples in the PCR [1].

1.2 Parasitism in free-living wild carnivorous mammals

Over the years, mammalian fauna has been declining more and more throughout the world for several reasons. Some of these factors are run over, the growing rapprochement between wild and domestic predators and breeding animals, the expansion of the agricultural frontier, formation of cattle pastures, and deforestation, which reduce natural environments, as well as increase environmental pollution, fur trade, and lack of prey in the natural environment [31, 32]. Another factor that can culminate in the diminution of this fauna is the parasitism by different etiological agents, like microorganisms, helminthes, and even arthropods, highlighting the gastrointestinal parasitosis. Wild mammals are constantly subjected to environmental conditions that favor the spread of parasites, even when restricted to restricted areas and populations [33].

The relationship between the environment, parasites, and hosts is extremely dynamic and has many equilibrium points that were reached during long periods of evolution [34]. The environment is the place that presents biotic and abiotic resources that allow the encounter, the survival, and maintenance of the life cycle of parasites and hosts. The parasites have the capacity to infect a large number and variety of hosts and, therefore, have important functions in the structuring of the communities, exerting great impact on the biodiversity and ecosystem dynamics [35]. Wild animals (hosts), on the other hand, present different degrees of susceptibility for a particular parasite and, thus, interfere both directly and indirectly in the success of parasitism by different etiological agents in ecosystems [35, 36].

The susceptibility of hosts and the ability of parasites to invade and colonize them are related to several factors, including the taxonomy, morphology, body size, and eating habits of the host in question [36]. Normally, species of taxonomically related hosts are susceptible to infections by the same species of parasites [37]. Therefore, the greater the taxonomic distance, the less likely that host parasites have characteristics compatible with other potential hosts [36]. Other aspects that interfere in the parasite-host relationship are the body size and the morphology of the animals. The thickness of the tegument, for example, and volume of the organs influence the invasion and survival capacity of the parasites in the host organism, and body size has great importance in the selection of foods to be eaten as well as in the place where the animals go hunting [38]. The feeding of the hosts has direct and indirect relation with the susceptibility of the same to the parasitic infections. Carnivorous diets are harmful to infections by intestinal protozoa, whereas herbivorous diets increase their potential for infection. In addition, plant-rich diets may exhibit antiparasitic effects [39].

Gastrointestinal parasites are one of the groups of agents that are transmitted and transmitted from one host to another in protected areas through predation, ingestion of water, or contact with contaminated soil. In general, helminths have been more reported than protozoa in free-living carnivores. This marked frequency of helminths in the different researches shows that the environment in which these animals circulate maintains favorable conditions for the maintenance of the cycle of nematodes, cestodes, trematoids, and acanthocephals, as well as the transmission of infective structures to these animals. The type of feeding ingested by the hosts can directly or indirectly affect the susceptibility of these animals to parasitic infections. Animals with meat-rich diets are more likely to have low prevalences of intestinal protozoa infections, whereas omnivorous or herbivorous diets increase the prevalence of infection by these agents. In addition, it is known that some plants that are ingested by animals may exhibit anthelmintic properties and the very friction of plant fiber may help omnivores and herbivores to purge helminth infections [36].

In Mexico, in tropical forests located in Veracruz and in the Yucatan Peninsula, 58.1% of positivity was reported in noninvasive fecal samples of wild felids identified using molecular techniques and with the aid of sniffing dogs. Two parasitological techniques were used, one of flotation and the other of spontaneous sedimentation [48, 49]. Among the parasites diagnosed were *Spirometra* sp. (33.5%), *Strongyloides* sp. (18%), *Physaloptera* sp. (11.4%), *Spirocerca* sp. (9%), Taeniidae (7.2%), phylum Acanthocephala (6.6%), *Ancylostoma* sp. (6.6%), *Toxocara* sp. (6%), ascarid-like eggs (3%), Coccidiasina oocysts (2.4%), *Capillaria* sp. (1.8%), *Gnathostoma* sp. (1.8%), *Uncinaria* sp. (1.8%), *Trichuris* sp. (1.2%), eggs of Anoplocephalidae (0.6%), and phylum Nematoda (0.6%). The authors observed that the parasitic communities of jaguar and puma were more similar between host species in the same forest type than among hosts inhabiting different forest types, which may have been influenced by the ecosystem differences and host evolutionary history, as well as disparate diet and habitat use of these two felines [40].

In Brazil, in Serra do Cipó National Park, 95% of positivity for gastrointestinal parasites was diagnosed in noninvasive fecal samples of *Chrysocyon brachyurus* and *Cerdocyon thous* identified macroscopically. The authors used three coproparasitological techniques, being two of spontaneous sedimentation [50, 51] and one of floatation [52]. The evolutionary forms detected in this study were mainly eggs of Trichuridae (68.4%). In addition, eggs of Ancylostomidae (52.6%), *Physaloptera* sp. (7.9%), Diphyllobothriidae (7.9%), Hymenolepidae (7.9%), *Toxocara* sp. (2.6%), Acanthocephala (2.6%), *Dipylidium caninum* (2.6%), *Isospora* sp. oocysts (2.6%), and *Strongyloides* sp. (2.6%) were also diagnosed. Despite the report of domestic dogs in Serra do Cipó National Park, signs of domestic dogs, such as feces, were found only in adjacent areas of the park where there are people communities. However, local residents reported seeing wild animals in the vicinity, indicating the possibility of a future proximity between these animals and perhaps their parasites [41].

After analyzing noninvasive fecal samples identified as felids by trichology in Serra dos Órgãos National Park, evolutionary forms of gastrointestinal parasites were detected in 88.6% of the feces analyzed using four different coproparasitological techniques, being two centrifuge floatations [53–55], one centrifuge sedimentation [51, 56], and one spontaneous sedimentation [57]. In this study, eggs of the Diphyllobothriidae family (65.8%) were the most detected parasites, followed by superfamilia Ascaridoidea (43.9%), nematode larvae (30.5%), Strongylidae order (21.9%), nonsporulated coccidian oocysts (9.8%), *Capillaria* sp. (7.3%), *Trichuris* sp. (6.1%), order Spirurida (4.9%), *Platynosomum* sp. (2.4%), and *Eimeria* sp. (1.2%). These results demonstrate that Serra dos Órgãos National Park presents all the elements necessary for maintenance of the biological cycles of different parasites, including those with complex biological cycles that include different types of hosts. Moreover, the laboratory diagnoses on the fecal samples enabled indirect partial analysis on the park ecosystem, being these stages of the parasites usually detected in free-living wild animals' samples, such as felids [42].

Gastrointestinal parasite infections can determine weight loss, metabolic imbalance, reproductive problems, anemia, and dehydration. In severe cases, they may also cause fetal malformations, locomotor lesions, and even death of the animal [43, 44]. Due to the clinical manifestations of gastrointestinal parasites, many hosts present behavioral and functional changes within their community. A predator at the top of the food chain, for example, may exhibit a decrease in food intake and activity, including hunting [35]. Furthermore, anthropic actions may result in the introduction of etiological agents, which in certain circumstances determine emerging infectious diseases in wild animal communities [45]. However, according to [46], populations of wild animals are generally in balance with their parasitological fauna. However, environmental changes, especially anthropogenic ones, can determine the introduction of new infective agents and even stress factors that destroy this equilibrium by inducing pathological situations. Parasites can be considered excellent bioindicators of environmental impacts [47]. In this way it becomes relevant to perform routine coproparasitological surveys with noninvasive samples collected in trails of conservation units in order to indirectly check the health of the environment.

2. Conclusions

The study of wild carnivores through the analysis of noninvasive samples allows the identification of the animal species by different techniques, as well as their monitoring without the exposure of these animals to situations of risk, stress, or the use of chemical tranquilizers by the researchers to manipulate the animals. In addition, through the analysis of noninvasive samples, mainly feces, it is possible to detect structures of gastrointestinal parasites that may potentially be infecting these animals. It should be emphasized that this type of sampling causes minimal interference in the carnivores' habitat during the collection of the biological samples in the environment by the researchers. Besides that, noninvasive sampling is not detrimental to the ecological niche of these animals, cooperating to maintain the integrity of the fauna and where they live. This sampling strategy is mainly important when studying regions' considered biodiversity hotspots, such as Madagascar Island, which has a unique and a high richness of biodiversity, which includes more than 98 species of mammals [58]. It is also important to highlight that the approaches about the biodiversity conservation have changed over the years, which means that conservation strategies are needed since all the species have their own function and values in the ecosystem, but also because they play a role in providing benefits to people and to economy, known as ecosystem services, producing food and materials, for example.

Therefore, the association of the results obtained from the identification of wild or domestic carnivores that share habitats, sympatric species, and the investigation of gastrointestinal parasites in the noninvasive biological samples of these animals are fundamental for understanding the effects of possible diseases that can affect wildlife. Moreover, a constant research of the gastrointestinal parasites in conservation units and protected areas is extremely important to detect possible human interference through the presence of specific parasites or the introduction of parasite taxa not commonly reported in free-living wild animals by invasive host species or domestic animals. In addition, noninvasive sampling is fundamental for updating the records on the circulation of wild fauna in conservation units, thus contributing to the creation or reformulation of management measures that aim, mainly, the preservation and perpetuation of these animal populations in the environment and also the non-entry and surveillance of animals that may be in the region and that do not belong to that ecosystem.

Acknowledgements

The authors thank the Laboratory of Parasitology of the Fluminense Federal University, the Itatiaia National Park and its employees, the Laboratory of Toxoplasmosis and Other Protozoan Diseases of the Oswaldo Cruz Institute, and CAPES.

Conflict of interest

There are no conflicts of interest.

Author details

Laís Verdan Dib[1]*, João Pedro Siqueira Palmer[1], Camila de Souza Carvalho Class de Lima[1], Otilio Machado Pereira Bastos[1], Claudia Maria Antunes Uchôa[1], Maria Regina Reis Amendoeira[2], Augusto Cezar Machado Pereira Bastos[1] and Alynne da Silva Barbosa[1,2]

1 Department of Microbiology and Parasitology, Biomedical Institute, Federal Fluminense University, Rio de Janeiro, Brazil

2 Oswaldo Cruz Foundation (Fiocruz, Rio de Janeiro), Toxoplasmosis and other Protozooses, Laboratory, Oswaldo Cruz Institute, Rio de Janeiro, Brazil

*Address all correspondence to: laisvdib@gmail.com

References

[1] Taberlet P, Waits LP, Luikart G. Noninvasive genetic sampling: Look before you leap. Trends in Ecology & Evolution. 1999;**14**(8):323-327

[2] Morin P, Chambers KE, Boesch C, Vigilant L. Quantitative polymerase chain reaction analysis of DNA from noninvasive samples for accurate microsatellite genotyping of wild chimpanzees (*Pan troglodytes verus*). Molecular Ecology. 2001;**10**(7):1835-1844

[3] Morin PA, Woodruff DS. Noninvasive genotyping for vertebrate. In: Smith TB, Wayne RK, editors. Molecular Genetic Approaches in Conservation. Oxford: University Press; 1996. pp. 298-313

[4] Chame M. Terrestrial mammal feces: A morphometric summary and description. Memórias do Instituto Oswaldo Cruz. 2003;**98**:71-94

[5] Bang P, Dahlström P. Huellas y Señales de los Animales de Europa. Omega: Barcelona; 1975

[6] Quadros J. Identificação microscópica de pêlos de mamíferos brasileiros e sua aplicação no estudo da dieta de carnívoros. Curitiba: Tese Doutorado Setor de Ciências Biológicas, Universidade Federal do Paraná; 2002

[7] Seton ET. On the study of scatology. Journal of Mammalogy. 1925;**6**:47-49

[8] Guandolini GC. Enriquecimento ambiental para gatos domésticos: A importância dos odores. Dissertação. Ribeirão Preto: Faculdade de Filosofia, Ciências e Letras, Universidade de Ribeirão; 2009

[9] Neilson J. Thinking outside the box: Feline elimination. Journal of Medicine and Surgery. 2004;**6**:5-11

[10] Ruiz F. Por que é que os gatos escondem suas fezes?. 2014. Disponível em: http://revistameupet.com.br/higiene-e-cuidados/porque-e-que-os-gatos-escondem-as-fezes/369/. Acess realized in: August 20th 2018

[11] Twigg GI. Finding mammals. Their sign and remains. Mammal Review. 1975;**5**:77-78

[12] Teerink BJ. Hair of West European Mammals: Atlas and Identification. Cambridge: University Press; 1991

[13] Hausman LA. Further studies of the relationships of the structural characters of mammalian hair. The American Naturalist. 1924;**58**(659):544-557

[14] Andrew W. A covering called skin. In: Textbook of Comparative Histology. Nova York: Oxford University Press, Oxford; 1959

[15] Day MG. Identification of hair and feather remains in the gut and faeces of stoats and weasels. Journal of Zoology. 1966;**148**:201-217

[16] Vanstreels RET, Ramalho FDP, Adania CH. Microestrutura de pelos-guarda de felídeos brasileiros: Considerações para a identificação de espécies. *Biota Neotropica*. 2010;**10**(1):333-337

[17] Dib LV. Pesquisa de parasitos gastrintestinais em felídeos silvestres do Parque Nacional da Serra dos Órgãos, RJ, Brasil. Monografia: Instituto Biomédico – Universidade Federal Fluminense; 2016

[18] Höss M, Kohn M, Pääbo S, Knauer F, Schröder W. Excrement analysis by PCR. *Nature*. 1992;**359**(6392):199

[19] Kohn MH, Wayne RK. Facts from feces revisited. Trends in Ecology & Evolution. 1997;**12**(6):223-227

[20] Rodgers TW, Janečka JE. Applications and techniques for non-invasive faecal genetics research in felid conservation. European Journal of Wildlife Research. 2013;**59**(1):1-16

[21] Farrell LE, Roman J, Sunquist ME. Dietary separation of sympatric carnivores identified by molecular analysis of scats. Molecular Ecology. 2000;**9**(10):1583-1590

[22] Avise JC. Phylogeography: Retrospect and prospect. Journal of Biogeography. 2009;**36**(1):3-15

[23] Kocher TD, Thomas WK, Meyer A, Edwards SV, Pääbo S, Villablanca FX, et al. Dynamics of mitochondrial DNA evolution in animals: Amplification and sequencing with conserved primers. Proceedings of the National Academy of Sciences. 1989;**86**(16):6196-6200

[24] Miotto RA, Cervini M, Kajin M, Begotti RA, Galetti PM. Estimating puma *Puma concolor* population size in a human-disturbed landscape in Brazil, using DNA mark–recapture data. Fauna & Flora International. 2014;**48**(2):250-257

[25] Folmer O, Black M, Hoeh M, Lutz R, Vrijenhoek R. DNA primers for amplification of mitochondrial cytochrome c oxidase subunit I from diverse metazoan invertebrates. Molecular Marine Biology and Biotechnology. 1994;**3**(5):294-299

[26] Hebert PD, Cywinska A, Ball SL. Biological identifications through DNA barcodes. Proceedings of the Royal Society of London B: Biological Sciences. 2003;**270**(1512):313-321

[27] Chaves PB, Graeff VG, Lion MB, Oliveira LR, Eizirik E. DNA barcoding meets molecular scatology: Short mtDNA sequences for standardized species assignment of carnivore noninvasive samples. Molecular Ecology Resources. 2012;**12**(1):18-35

[28] Rodríguez-Castro KG, Saranholi BH, Bataglia L, Blanck DV, Galetti PM Jr. Molecular species identification of scat samples of south American felids and canids. Conservation Genetics Resources. 2018;**10**:1-6

[29] Davison A, Birks JD, Brookes RC, Braithwaite TC, Messenger JE. On the origin of faeces: Morphological *versus* molecular methods for surveying rare carnivores from their scats. Journal of Zoology. 2002;**257**(2):141-143

[30] Graeff VG. Identificação de espécies de carnívoros brasileiros (Mammalia: Carnivora) a partir de amostras de fezes utilizando sequências de DNA e microscopia óptica de pelos. Dissertação. Vol. 2008. Faculdade de Biociências - Pontifícia Universidade Católica do Rio grande do Sul; 2008

[31] Wilson DE, Mittermeier RA. The Mammals of the World. Lynx: Barcelona; 2009

[32] Aranda C, Serrano-Martínez E, Tantaleán M, Quispe M, Casas G. Identificación y frecuencia de parásitos gastrointestinales en félidos silvestres en cautiverio en el Perú. Revista de Investigaciones Veterinarias del Perú. 2013;**24**(3):360-368

[33] Mangini PR, Vidolin GP, Velastin GO. Pesquisa de macroparasitos em carnívoros selvagens: uma ferramenta para a conservação. In: MORATO et al., editors. Manejo e conservação de carnívoros neotropicais. São Paulo: IBAMA; 2006. pp. 307-323

[34] Araujo A, Serrano-Martínez E, Tantaleán MV, Quispe MH, Casas GV. Parasitism, the diversity of life, and paleoparasitology. Memórias do Instituto Oswaldo Cruz. 2003;**98**:5-11

[35] Poulin R. The functional importance of parasites in animal communities: Many roles at many

levels? International Journal for Parasitology. 1999;**29**(6):903-914

[36] Freeland WJ. Parasites and the coexistence of animal host species. American Naturalist. 1983;**21**:223-236

[37] Stunkard HW. Host-specificity and parallel evolution of parasitic flatworms. Zeitschrift für Tropenmedizin und Parasitologie. 1957;**8**:254-263

[38] Vinson SB. Behavioral chemicals in the augmentation of natural enemies. In: Ridgway RL, Vinson SB, editors. Biological Control by Augmentation of Natural Enemies: Insect and Mite Control with Parasite and Predators. Nova York: Plenum; 1977. pp. 237-279

[39] Hegner RW. The relations between a carnivorous diet and mammalian infection with intestinal protozoa. American Journal of Epidemiology. 1924;**40**:393-400

[40] Solórzano-García B, White-Day JM, Contreras MG, Cristóbal-Azkáratec J, Osorio-Sarabiad D, Rodríguez-Luna E. Coprological survey of parasites of free-ranging jaguar (*Panthera onca*) and puma (*Puma concolor*) inhabiting 2 types of tropical forests in Mexico. Revista Mexicana de Biodiversidad. 2017;**88**(1):146-153

[41] Santos JLC, Magalhães NB, Santos HÁ, Ribeiro RRR, Guimarães MP. Parasites of domestic and wild canids in the region of Serra do Cipó National Park. Brazil. Revista Brasileira de Parasitologia Veterinária. 2012;**21**(3):270-277

[42] Dib LV, Cronemberger C, Pereira FA, Bolais PF, Uchôa CMA, Bastos OMP, et al. Gastrointestinal parasites among felids inhabiting the Serra dos Órgãos National Park, Rio de Janeiro. Brazil. Revista Brasileira de Parasitologia Veterinária. 2018;**27**(2):131-140

[43] Azipiri GS, Maldonado FG, González GC. La importancia del estudio de enfermedades en la conservación de fauna silvestre. Veterinária. México. 2000;**31**(3):223-230

[44] Barutzki D, Schaper R. Endoparasites in dogs and cats in Germany 1999-2002. Parasitology Research. 2003;**90**(3):S148-S150

[45] Daszak P, Cunningham AA, Hyatt AD. Emerging infectious diseases of wildlife and threats to biodiversity and human health. Science. 2000;**287**(5452):443-449

[46] Sousa CB, Madeira de Carvalho L, Fazendeiro I, Castro Rego F, Afonso-Roque M. Contribution for the knowledge of wild boar (*Sus scrofa* L.) helmintic fauna in Tapada Nacional de Mafra, an enclosured hunting area. Revista Ibérica Parasitológica. 2004;**64**:3-7

[47] Vidal-Martinez VM, Pech D, Sures B, Purucker ST, Poulin R. Can parasites really reveal environmental impact? Trends in Parasitology. 2010;**26**(1):44-51

[48] Greiner EC, McIntosh A. Collection methods and diagnostic procedures for primate parasitology. In: Huffman MA, Chapman CA, editors. Primates Parasite Ecology. The Dynamics and Study of Host–Parasite Relationships. Cambridge: Cambridge University Press; 2009

[49] Hendrix CM. Diagnostic Veterinary Parasitology. 2nd ed. St. Louis, Missouri: Mosby; 1998

[50] Hoffman WA, Pons JA, Janer JL. The sedimentation-concentration method in Schistosomiasis mansoni. The Puerto Rico Journal of Public Health and Tropical Medicine. 1934;**9**:283-291

[51] Ritchie LS. An ether sedimentation technique for routine stool

examinations. U.S. Army Medical Department Journal. 1948;**8**(4):326

[52] Willis HH. A simple levitation method for the detection of hookworm ova. Medical Journal of Australia. 1921;**11**:375-376

[53] Faust EC, D'Antoni JS, Odon V, Miller MJ, Perez C, Sawitz W, et al. A critical study of clinical laboratory technics for the diagnosis of protozoan cysts and helminth eggs in feces. I. Preliminary communication. American Journal Tropical Medicine. 1938;**18**:169-183

[54] Sheather AL. The detection of intestinal protozoa and mange parasites by a flotation technique. The Journal of Comparative Pathology and Therapeutics. 1923;**36**:266-275

[55] Huber F, Bonfim TC, Gomes RS. Comparação da eficiência da técnica de sedimentação pelo formaldeído-éter e da técnica de centrífugo-flutuação modificada na detecção de cistos de *Giardia* sp. e oocistos de *Cryptosporidium* sp. em amostras fecais de bezerros. Revista Brasileira de Parasitologia Veterinária. 2003;**12**(2):135-137

[56] Young KH, Bullock SL, Melvin DM, Spruill CL. Ethyl acetate as a substitute for diethyl ether in the formalin-ether sedimentation technique. Journal of Clinical Microbiology. 1979;**10**:852-853

[57] Lutz A. O *Schistosomum mansoni* e a schistosomatose segundo observações feitas no Brasil. Meméмorias do Instituto Oswaldo Cruz. 1919;**11**(1):121-155

[58] Ganzhorn JU, Lowry PP II, Schatz GE, Sommer S. The biodiversity of Madagascar: One of the world's hottest hotspots on its way out. Oryx. 2001;**35**(4):346-348